EASY AI FOR BUSINESS MANAGEMENT

A PRACTICAL GUIDE TO SMARTER DECISIONS, LEANER OPERATIONS, AND FUTURE-READY TEAMS

MICHAEL GRANT MALLOY

CONTENTS

AUTHOR'S NOTE

When people hear the phrase *Artificial Intelligence*, they often imagine futuristic robots or complex algorithms they could never understand. The truth is, AI is already woven into the way we shop, work, and live. From recommending the next show you'll watch to flagging unusual activity on your credit card, AI is working quietly in the background — and it's time for business leaders to bring it into the foreground.

This book was written for business leaders, managers, and professionals who want to cut through the hype and focus on what AI can do for their organizations. You won't find equations or programming tutorials here. Instead, you'll find plain-language explanations, real-world case studies, and actionable frameworks designed to help you make smarter decisions, run leaner operations, and prepare your teams for the future. You'll also find examples of current AI platforms on the market that may be just what your business is looking for.

My belief is simple: AI is not about replacing human intelligence but amplifying it. It helps us see patterns we would otherwise miss, respond to challenges before they become disasters, and free up time for the creative, strategic, and human work that matters most. Let's make AI work for your organization - clearly, confidently, and effectively.

PREFACE

Artificial intelligence is no longer the stuff of science fiction or the exclusive domain of tech giants. It has seamlessly integrated into everyday business practices, influencing everything from customer service to supply chain management and strategic decision-making. And yet, for many business leaders, AI still feels complicated, intimidating, or as if it's reserved for companies with armies of data scientists and billion-dollar budgets.

This book was written to break down that barrier. My goal is simple: to make AI accessible, understandable, and actionable for professionals without a technical background, who are eager to leverage AI's potential for business advantage. You don't need to know how to code or build algorithms from scratch. What you do need is a framework for understanding what AI can do, where it adds value, and how to implement the power of AI responsibly for your organization.

Over the past decade, I've watched organizations wrestle with digital transformation in various ways. What separates those who succeed from those who stall is not technical prowess. Instead, it's clarity of purpose, a willingness to start small, and the discipline to scale what works. AI tools, when applied thoughtfully, strengthen strategic planning, streamlines daily operations, and unlocks opportunities for growth that weren't possible even a few years ago. This book offers a roadmap to help you on that journey.

The chapters that follow will provide insights into how AI can impact essential domains of business management where AI is already making a measurable impact: finance, workforce productivity, customer experience, marketing, supply chain, and beyond. Each chapter blends explanation with real-world case studies, showing how companies of all sizes are deploying AI today. Along the way, you'll find guiding questions, checklists, and templates designed to help you apply these lessons to your own context.

Importantly, this is not a book about replacing people with machines. Quite the opposite. At its core, AI is most powerful when it augments human judgment, freeing managers and teams from repetitive tasks so they can focus on higher-value work. The future belongs to leaders who can blend human creativity and empathy with AI-driven insights.

My hope is that by the time you finish this book, AI will feel less like a mysterious buzzword and more like a practical set of tools you can put to work immediately. Whether you lead a team of ten or a company of ten thousand, you'll walk away with the confidence to ask better questions, choose smarter tools, and guide your organization into an AI-enabled future. Welcome to *Easy AI for Business Management*. Let's make AI work for you — simply, practically, and effectively.

CHAPTER 1
INTRODUCTION: WHY AI MATTERS FOR BUSINESS MANAGEMENT

IF YOU'VE BEEN in business for more than a decade, you've probably seen technology reshape the way we work. Email replaced faxes. Cloud computing replaced filing cabinets. Smartphones put a miniature office in everyone's pocket. Each shift forced leaders to adapt—or risk being left behind.

Artificial intelligence (AI) is the next leap, but unlike earlier tools, it isn't just about making processes digital or mobile. It's about making them intelligent. AI doesn't just automate tasks; it analyzes, predicts, learns, and in some cases, recommends better paths forward. That makes it less of a "new tool" and more of a new partner in the way businesses are run.

From Hype to Practical Value

It's easy to dismiss AI as another overhyped tech buzzword. After all, you've probably read headlines about self-driving cars or AI art, which may feel far removed from managing budgets, directing teams, or building customer relationships. But here's the reality: AI is already working behind the scenes in many of the tools you use every day.

- When your email filters out spam, that's AI.

- When your GPS predicts the fastest route during rush hour, that's AI.
- When your bank flags a suspicious transaction before you even notice, that's AI.

The same intelligence that powers these conveniences is now available to businesses of every size, not just Fortune 500 companies with massive R&D budgets. Cloud-based AI platforms, affordable subscription tools, and built-in AI features are democratizing access. In other words: AI isn't just for tech giants anymore—it's for you.

Why Mid-Career Leaders Can't Ignore AI

If you're reading this, you're likely mid-career. You're experienced enough to have weathered multiple waves of technology change, but also seasoned enough to know that leadership today isn't about doing everything yourself. It's about making smart choices with limited resources.

Here's why AI should be on your radar now:

1. **Competitive Advantage** – Businesses that adopt AI early tend to move faster, serve customers better, and operate leaner than those that don't. Your competitors are already experimenting with it—some quietly, some loudly.
2. **Efficiency Gains** – AI can handle repetitive tasks (like scheduling, invoicing, or initial customer support) so your team can focus on higher-value work. Think of it as hiring an assistant who works 24/7, never asks for vacation, and costs less than a single salary.
3. **Smarter Decisions** – Leaders are drowning in data but starving for insights. AI helps turn noise into clarity—showing you patterns in sales, risks in supply chains, or hidden drivers of customer loyalty.
4. **Talent Pressures** – Hiring is harder and more expensive than ever. AI tools can help attract, evaluate, and retain talent by handling early-stage screening, monitoring employee sentiment, and even recommending reskilling opportunities.

5. **Future-Proofing** – Whether we like it or not, AI is becoming the "new normal." Ignoring it is like ignoring email in the 1990s. It won't stop the world from moving forward. It will just make you irrelevant faster.

Shifting the Mindset: From Threat to Opportunity

One of the biggest barriers to AI adoption isn't technology. One of the biggest barriers is mindset. Many professionals worry: *Will AI replace me? Will my years of experience still matter? Will I be able to adapt?* These are valid concerns, but they tend to frame AI as a threat rather than an ally.

The truth is, AI doesn't erase experience. AI amplifies experience. An AI tool can crunch numbers at lightning speed, but it can't navigate office politics, inspire a team, or build trust with a new customer. Those are human strengths. By letting AI handle the heavy lifting of data and routine tasks, you free up time and energy to lean into the leadership skills that machines can't replicate.

Think of AI less like a robot competitor and more like a junior partner. A partner that works tirelessly, makes useful suggestions, and helps you see what you might miss. The leaders who thrive will be those who learn to collaborate with AI, not compete with it.

Real-World Proof: Businesses Already Winning with AI

To ground this in reality, let's look at a few examples:

- **Starbucks** uses AI to analyze customer purchase histories and local weather patterns, tailoring offers through its app. This keeps customers engaged and drives repeat business.
- **UPS** relies on AI to optimize delivery routes, saving millions of gallons of fuel each year while simultaneously improving delivery times.
- **Small retailers** now use AI-powered inventory tools that predict which products will sell best and during which season, allowing them to compete with giants like Amazon.

Notice something? None of these are sci-fi applications. They're practical, operational, and profitable.

Your Journey Through This Book

This book is not intended to turn you into a programmer or data scientist. Instead, it aims to equip you with the knowledge and confidence to utilize AI as a management tool, similar to how previous generations embraced the printing press, the assembly line, the internet and email.

Here's what's ahead:

- We'll explore how AI can make strategic planning more informed.
- We'll dive into finance, operations, and workforce management, where AI can eliminate inefficiencies.
- We'll look at marketing, customer service, and decision-making, where AI gives leaders a sharper edge.
- And we'll tackle the risks and ethics, because responsible adoption is just as important as ambitious adoption.

By the end, you should understand not just why AI matters, but how to apply AI to your own role, your team, and your organization.

Closing Thought

Every major shift in business creates winners and losers. The winners aren't always the ones with the most resources. The winners are the ones who adapt fastest. AI is not the future of business management. It is the present. It is here now. The only question left for you is: Will you lead the change, or wait for it to leave you behind?

CHAPTER 2
STRATEGIC PLANNING WITH AI

IF YOU'VE EVER SAT in a strategy meeting staring at slides full of market forecasts, you know how hard it is to plan for a future that keeps changing. The traditional approach to strategy—analyzing last year's numbers, making cautious projections, and crossing your fingers—feels increasingly inadequate in a world where customer preferences, technology, and even entire industries can shift overnight.

This is where artificial intelligence (AI) becomes a game-changer. AI doesn't just look backward; it looks into the future. AI doesn't just crunch numbers; it finds patterns or trends in those numbers you didn't know existed. AI doesn't just recommend the "most likely" outcome, it can simulate multiple scenarios and suggest paths that help you anticipate market changes in real time instead of merely reacting to them.

From Gut Instinct to Data-Driven Foresight

Mid-career business professionals often rely on a mix of experience, intuition, and trend reports to guide their decisions. Those instincts are valuable, but even the sharpest leaders can only process so much information at once. AI, by contrast, thrives on complexity. AI can ingest data from hundreds of sources, such as customer transactions, social media chatter, economic indicators, competitor activity. It can then turn that data into a clear map of possible futures.

Imagine being able to ask: *What if inflation rises by 3% next quarter? How would that affect my customer demand, supply chain, and hiring needs?* AI can model those scenarios in real time, giving you insights that, in the past, would have taken your team weeks of painstaking analysis.

How AI Strengthens Strategic Planning

When most leaders hear the term "strategic planning," they think of long off-site meetings, thick binders of reports, and charts projecting growth curves that may or may not come true. The problem with strategic planning in the past wasn't a lack of effort. The problem was that traditional strategic planning relied heavily on historical data and human interpretation, which can only go so far in an unpredictable world. AI changed that by bringing speed, scale, and foresight into the strategic planning process. Let's look at four ways AI strengthens strategic planning:

1. Scenario Forecasting

One of AI's most powerful contributions is its ability to model multiple possible futures. Instead of building a single forecast that assumes "business as usual," AI can generate dozens of "what-if" scenarios. For example: *What happens to sales if inflation rises another 2%? What if a new competitor enters the market? What if supply costs drop by 10%?* By analyzing millions of data points and running complex simulations, AI allows leaders to plan for uncertainty, not just predictability. The result? More resilient strategies that don't collapse the moment reality veers from the plan.

2. Competitive Intelligence

In the past, keeping tabs on competitors meant scanning trade publications, waiting for quarterly reports, or hearing rumors from customers and suppliers. Today, AI tools can scan thousands of data sources—news articles, job postings, patent filings, press releases, even customer reviews—and piece together a picture of what your rivals are planning. If your competitor is quietly hiring data scientists or filing patents in a new area, AI can flag those signals long before a new product hits the market. That gives you a valuable head start in adjusting your own strategy.

3. Customer Trend Prediction

Customer behavior often shifts faster than businesses can track. AI excels at identifying patterns that humans miss, especially in consumer-facing industries. By analyzing purchase histories, search queries, and even social media sentiment, AI can forecast emerging preferences. Imagine being able to spot that your customers are leaning toward eco-friendly packaging or subscription models before it becomes mainstream. That kind of foresight enables you to adapt your offerings before competitors catch up, positioning your business as the trendsetter rather than the follower.

4. Resource Prioritization

At its core, strategy is about choices: what to pursue and what to leave behind. Yet many companies spread themselves too thin, chasing every opportunity. AI helps leaders make smarter choices by predicting which initiatives are most likely to succeed. By analyzing market demand, cost structures, and historical performance, AI can rank projects based on potential ROI. This doesn't mean you blindly follow the algorithm—but it gives you a powerful filter to separate shiny distractions from meaningful growth opportunities.

In short, AI doesn't eliminate the need for leadership. AI equips leaders with a clearer, broader lens. Instead of reacting to surprises, you gain the ability to anticipate, prioritize, and move decisively. For mid-career professionals, this means your experience is amplified by data-driven foresight, making your strategic decisions sharper, faster, and more credible.

Case Studies – AI in Action

Talking about AI in theory is one thing, but seeing AI in practice makes the value far clearer. Across industries, companies both large and small are already using AI to strengthen their strategic planning. These real-world examples show how practical AI has become.

- **Shell Energy: Planning for Volatile Futures**

The energy sector is notoriously unpredictable. Oil prices swing with geopolitics, renewable energy adoption is accelerating, and regulations change by region. Shell uses AI to model multiple global demand scenarios, integrating data from climate forecasts, regulatory trends, and consumer adoption rates. Instead of planning around a single projection, Shell's executives can explore dozens of potential futures and decide how to allocate investments accordingly. This strategic flexibility helps them avoid costly missteps while positioning themselves to pivot quickly.

- **Retail Fashion: Predicting What Customers Will Want**

Fashion retailers live and die by their ability to predict trends. Traditionally, designers and buyers relied on past sales data, intuition, and fashion weeks. Now, AI is changing the game. Companies use AI to scan millions of Instagram posts, TikTok videos, and e-commerce searches to identify which styles, colors, or fabrics are gaining momentum. One European fashion brand discovered through AI analysis that pastel tones were surging months before the competition caught on, allowing them to launch collections ahead of the curve and capture outsized market share.

- **Logistics and Expansion Decisions**

Mid-sized logistics companies often face a risky question: Should we expand into new regions? The wrong move can waste millions on warehouses, trucks, and staff. Using AI, some firms model regional demand, fuel costs, and competitor density to simulate outcomes before committing resources. In one case, a U.S. regional carrier avoided a costly expansion into an area where AI predicted low profitability and instead redirected funds into optimizing its existing routes —improving margins without overextending.

- **Healthcare Systems: Managing Resource Allocation**

Hospitals and healthcare systems must balance patient demand with limited staff and resources. AI is now being used to forecast seasonal surges (like flu spikes) and allocate staff accordingly. For example, one hospital system used AI to predict emergency room traffic weeks in advance, reducing wait times and improving patient satisfaction scores. This wasn't just an operational win—it became a core element of the hospital's strategic planning, helping them allocate budgets more effectively and improve overall care quality.

- **Small Business Applications**

AI isn't only for multinationals. Small retailers and e-commerce shops are using AI-driven tools to plan inventory, manage digital ad spending, and optimize pricing strategies. A boutique in California used AI to identify which products would likely sell out and which would stagnate, and were able to adjust their orders accordingly. The result? Less money tied up in unsold inventory and more cash freed for marketing.

The lesson from these case studies is simple: AI isn't a distant promise. AI is a present-day tool that organizations of all sizes can leverage. Whether it's optimizing inventory, predicting consumer behavior, or modeling global scenarios, AI helps leaders shift from guesswork to foresight. And in a business environment where uncertainty is the only constant, foresight is priceless.

Closing Thought

Strategic planning has always been about balancing today's reality with tomorrow's possibilities. The difference now is that you don't have to do it blindfolded. AI is like having a radar system for your business—it won't make the decisions for you, but it will reveal the storms and opportunities on the horizon.

For mid-career leaders, this is especially powerful. You already bring years of industry wisdom and managerial experience. Pair that with AI-driven foresight, and you have the best of both worlds: the intuition of a seasoned professional backed by the predictive power of intelligent technology.

📌 Guiding Questions for Leaders

To make this discussion practical, ask yourself:

- Do I currently plan for multiple possible futures, or do I bet on just one?
- How well do I understand what my competitors are doing right now?
- What "early warning signals" am I missing because my team doesn't have time to sift through mountains of data?
- Where could AI free me from guesswork and give me confidence in my choices?

CHAPTER 3
WHY DATA QUALITY DETERMINES AI SUCCESS

THERE'S an old saying in computer science: "garbage in, garbage out." Nowhere is that truer than in artificial intelligence. No matter how sophisticated an AI algorithm is, if it's fed incomplete, inconsistent, or biased data, the results will be unreliable at best — and damaging at worst. Think of AI as the engine of a modern business. That engine doesn't run on ideas, vision, or even capital alone. It runs on data. Clean, structured, and relevant data is the fuel that determines whether AI delivers real insights or misleading noise. Without it, companies risk building expensive AI systems that fail to generate value.

The stakes are high. Poor data quality can cost organizations millions per year in wasted resources, inefficiency, and lost opportunities. Meanwhile, a recent Harvard Business Review study found that 47% of newly created data records contain at least one critical error, such as a missing address, incorrect spelling, or outdated information (Redman, 2018). Errors like these don't just frustrate employees, they ripple through AI systems, amplifying bad decisions at scale and draining the bottom line.

Clean, structured data enables AI to move beyond descriptive analytics ("what happened") toward predictive and prescriptive analytics ("what's likely to happen next, and what should we do about it?"). For example, a retail chain using AI to optimize inventory relies on accurate sales data, complete product identifiers, and standardized location codes. If that data is wrong or incomplete, the AI may recommend sending the wrong products to the wrong stores — undermining both profitability and customer trust.

On the flip side, organizations that treat data as a strategic asset unlock transformative possibilities. When Starbucks structured and centralized customer data from its loyalty app, POS systems, and mobile orders, it gave its AI recommendation engine the ability to personalize promotions at scale. As a result, Starbucks increased revenue per customer by tailoring offers in real time based on purchase patterns and preferences (Starbucks, 2019).

The lesson is clear: before investing in AI tools, businesses must ask themselves if their data is ready for AI. Success is not about having the fanciest AI algorithm — it's about having the cleanest, most trustworthy dataset powering their AI algorithm. AI doesn't magically fix bad data. In fact, AI will likely magnify bad data through bad outcomes.

📌 Guiding Questions for Reflection

- Do we know how much of our business data is clean, consistent, and up to date?
- Are we tracking the hidden costs of poor data quality — in missed opportunities or customer dissatisfaction?
- Is our organization treating data as a strategic asset — or just as digital exhaust?

The Anatomy of Business Data

To understand how to prepare your organization for AI, it helps to know what kinds of data your business generates, and how different types of data fuels different AI applications.

Broadly, business data falls into three categories: structured, unstructured, and external.

1. Structured Data: The "Orderly" Foundation

Structured data is what most businesses are familiar with. It lives in databases, spreadsheets, and enterprise systems like CRM (customer relationship management) and ERP (enterprise resource planning) platforms.

Examples include:

- Customer names, phone numbers, and email addresses.
- Sales transactions, SKUs, and inventory levels.
- Financial data such as revenue, expenses, and forecasts.

This data is highly organized — think of rows and columns in a spreadsheet. Structured data is relatively easy for AI to process, making it the backbone of applications like predictive sales forecasting, inventory optimization, and fraud detection.

For instance, in banking, AI fraud-detection models analyze millions of structured transaction records in real time, flagging anomalies that deviate from normal customer behavior.

2. Unstructured Data: The Untapped Goldmine

Unstructured data doesn't fit neatly into rows and columns. It makes up about 80–90% of the world's data, yet much of it remains under-utilized.

Examples include:

- Customer reviews on Amazon or Yelp.
- Emails, support chat transcripts, and call recordings.
- Social media posts, images, and videos.
- IoT sensor data (data collected by sensors embedded into physical things, such as car engines), such as machine sounds or temperature readings.

AI, particularly natural language processing (NLP) and computer vision, unlocks value from this messy, text-heavy, and multimedia-rich data. For example, airlines use NLP to analyze customer service calls, identifying frustration or satisfaction levels in real time. Retailers analyze product reviews to detect emerging issues before they escalate.

The challenge is that unstructured data is harder to clean, label, and interpret. But once harnessed, it can provide richer, more human insights than structured data alone.

3. External Data: Context that Complements

No business operates in a vacuum. External data sources often provide the context that makes predictions and insights more accurate.

Examples include:

- Weather forecasts influencing retail demand or shipping delays.
- Social media sentiment shaping stock price predictions.
- Market trend data guiding product launches.
- Macroeconomic indicators (GDP growth, inflation) feeding into financial forecasts.

For example, some big box stores integrate weather data into its demand forecasting models. A predicted cold front can trigger higher stock levels of heaters and winter apparel in regional stores, a decision powered by external signals combined with internal sales data.

Bringing It All Together

AI is most powerful when it combines these three types of data. Structured data provides the foundation, unstructured data enriches understanding, and external data supplies critical context. Think of it like this:

- Structured data tells you *what happened* (sales dropped 5% last week).
- Unstructured data tells you *why* (customers complained about poor product quality on social media).

- External data tells you *what might happen next* (a competitor is trending online, or a storm is likely to affect shipping routes).

The businesses that thrive in the AI era will be those that not only collect these types of data, but also integrate them into a unified view of the customer, the market, and the operation.

📌 Guiding Questions for Reflection

- Do we have visibility into all three types of data: structured, unstructured, and external?
- Which of these data sources are we currently underutilizing?
- Could integrating external signals make our AI predictions more accurate and actionable?

Data Challenges in Business Contexts

If data is the fuel for AI, then many businesses are trying to run high-performance engines on contaminated, fragmented, or incomplete fuel supplies. While executives are eager to harness AI, most organizations still face significant hurdles in preparing their data for success.

1. Data Silos Across Departments

One of the most common challenges is data silos, which is when different departments, marketing, sales, finance, HR, operations, each maintain their own separate databases. These databases aren't set up to "talk" to each other, making it nearly impossible to build a unified picture of customers or operations.

For example, a retailer might have purchase history in its POS system, customer loyalty data in its CRM, and supply data in its ERP. Without integration, AI models are forced to work with fragmented inputs, producing incomplete or contradictory insights or outputs.

2. Inconsistent and Incomplete Data

Dirty or incomplete data is another major barrier. Common problems include:

- Missing customer contact information.
- Duplicate records for the same individual.
- Outdated entries (e.g., addresses that haven't been updated for years).
- Inconsistent formats (e.g., dates entered differently across systems).

A Harvard Business Review study found that 47% of newly created data records contain at least one critical error (Redman, 2018). When fed into AI, these inconsistencies don't just stay local — they ripple across predictive models, producing inaccurate recommendations or flawed forecasts.

3. Legacy Systems and Integration Barriers

Many organizations still rely on decades-old legacy IT systems. These systems weren't designed for today's volumes of data, or for integration with modern AI platforms. Extracting data from these outdated systems can be slow, expensive, or technically challenging, and the data is often incomplete. Unfortunately, replacing old systems can be cost prohibitive. This dilemma creates a paradox: businesses want AI to modernize their operations, but outdated infrastructure makes that modernization process harder.

4. Privacy and Compliance Constraints

The explosion of privacy regulations — from Europe's General Data Protection Regulation (GDPR) to California's California Consumer Privacy Act (CCPA) — means businesses must handle customer data with far greater care. Collecting and using data without clear consent, or failing to protect it, can result in fines, reputational damage, and loss of customer trust. AI initiatives often stumble here. Companies may have rich datasets, but if those datasets can't be used in compliance with regulations, AI project often stall. Combine these constraints with outdated IT systems, and instead of modernizing operations, attempts at implementing AI systems slow the organization down even further.

GDPR vs. CCPA Comparison

Aspect	GDPR (General Data Protection Regulation)	CCPA (California Consumer Privacy Act)
Jurisdiction	European Union (applies to all EU member states, with extraterritorial reach worldwide if handling EU citizens' data)	State of California, USA (applies to businesses handling California residents' data)
Effective Date	May 25, 2018	January 1, 2020
Who Must Comply	Any organization processing personal data of EU residents, regardless of location	For-profit businesses meeting thresholds (>$25M revenue, 50,000+ consumers' data, or >50% revenue from selling data)
Definition of Personal Data	Broad: any information relating to an identified or identifiable person (e.g., name, ID, IP address, location, biometric data)	Slightly narrower: information that identifies, relates to, describes, or can reasonably be linked to a consumer or household
Key Consumer Rights	- Right to access - Right to rectification - Right to erasure ("right to be forgotten") - Right to restrict processing - Right to data portability - Right to object to processing	- Right to know what data is collected - Right to delete personal data - Right to opt out of sale of personal data - Right to non-discrimination for exercising privacy rights
Consent Model	Opt-in: explicit consent required for most data processing activities	Opt-out: consumers must take action to stop the sale of their data
Penalties	Up to €20 million or 4% of annual global turnover, whichever is higher	Up to $7,500 per intentional violation; $2,500 per unintentional violation
Enforcement Authority	Supervisory authorities in each EU member state	California Attorney General (and California Privacy Protection Agency since 2021)
Global Impact	Became the gold standard for global privacy laws; influenced legislation worldwide	Inspired other U.S. states (e.g., Virginia, Colorado, Connecticut, Utah) to introduce similar privacy laws

5. Cultural Barriers: Not Treating Data as an Asset

Perhaps the most subtle, but most damaging, barrier is cultural. In many companies, employees don't see data quality as their responsibility. Salespeople cut corners when entering CRM records. Call center staff fail to log interactions completely. There are innumerable ways that an organizations data can be corrupted from within. Without a culture that values data accuracy, errors accumulate and AI performance for the organization falls short.

As Thomas Redman, a leading expert on data quality, puts it: *"If your people aren't committed to managing data as a strategic asset, no technology will save you."* (Redman, 2018).

The Cost of Ignoring These Challenges

When data is siloed, inconsistent, outdated, or non-compliant, AI initiatives are destined to struggle. Gartner estimates that 85% of AI projects that fail to deliver business value are often due to poor data foundations (Gartner, 2019). In other words, the challenges above don't just slow an organizations progress, they can derail it entirely.

📌 **Guiding Questions for Reflection**

- Are we still operating with data silos that prevent a unified view of customers or operations?
- How much of our data is "dirty" — incomplete, inconsistent, or outdated?
- Are our IT systems modern enough to support large-scale data integration?
- Do employees treat data as a strategic asset, or just an afterthought?

Preparing Data for AI Readiness

If data is the fuel of AI, then preparation is the refinery. Without refining, raw data often contains too much noise, inconsistency, and fragmentation to be useful. Preparing your business data for AI readiness requires a systematic approach — one that combines technical processes with organizational discipline.

1. Data Auditing: Knowing What You Have

The first step is data discovery — mapping where data resides, what forms it takes, and how it flows through your business. This process often reveals hidden silos, duplicate repositories, or underutilized datasets.

A data audit should answer:

- What systems store our customer, financial, and operational data?
- How complete and accurate is each dataset?
- Who owns and maintains the data?

Without this visibility, AI projects are like building a house without knowing what materials are available. Studies show that organizations with formal data governance and audit practices achieve better AI outcomes and reduce compliance risks (Khatri & Brown, 2010).

2. Data Cleaning: Eliminating "Dirty Data"

Data cleaning, also called "data scrubbing," addresses inaccuracies, inconsistencies, and missing values. Common activities include:

- Removing duplicate records.
- Correcting formatting inconsistencies (e.g., date formats, units of measurement).
- Filling in missing fields or validating against external reference sources.
- Standardizing naming conventions across systems.

The ROI is tangible. Experian found that companies lose an average of 12% of revenue due to inaccurate data, mostly through wasted marketing and missed opportunities (Experian, 2021). Clean data not only improves AI outcomes, it also enhances everyday decision-making.

3. Data Integration: Breaking Down Silos

Once data is clean, it must be connected. Integration involves linking Customer Relationship Management (CRM), Enterprise Resource Planning (ERP), Human Resources, marketing, and supply chain data into a unified view.

Techniques include:

- Extract, Transform, Load (ETL) pipelines to consolidate data into warehouses.
- Application Programming Interfaces (API) that connect systems in real time.
- Master Data Management (MDM) to ensure consistent records across platforms.

A unified dataset enables AI to recognize patterns across departments. For example, combining customer service tickets (unstructured) with sales history (structured) allows AI to predict churn more accurately.

4. Metadata and Data Labeling: Adding Context

AI models rely on context to make sense of data. That context comes from metadata (descriptions of the data) and labels (used for training supervised learning).

For example, labeling images of defective vs. non-defective products enables a computer vision model to detect flaws in manufacturing. Metadata such as "last updated date" ensures models aren't trained on stale information.

Neglecting this step can cripple AI initiatives. In fact, labeling and annotation often account for up to 80% of the effort in machine learning projects (Zhou et al., 2017).

5. Data Governance: Setting Standards and Accountability

Data readiness isn't a one-time project, it's an ongoing discipline. Data governance provides the policies, processes, and roles that ensure data remains accurate, consistent, and secure over time.

Best practices include:

- Defining clear data ownership and stewardship.
- Establishing standards for accuracy, completeness, and timeliness.

- Implementing access controls to safeguard sensitive information.
- Creating escalation processes for correcting errors.

Effective governance turns data management from a back-office chore into a strategic capability. Organizations with strong governance frameworks are far more likely to trust and adopt AI-driven recommendations.

Bringing It Together: From Raw Data to AI-Ready Assets

Preparing data for AI is not glamorous, but it's indispensable. It requires technical rigor (cleaning, integration, labeling) and cultural commitment (governance, accountability). Companies that master this preparation stage dramatically increase their chances of AI success. In other words, before asking *"What can AI do for us?"* organizations should ask *"What can we do to make our data AI-ready?"*

🖈 **Guiding Questions for Reflection**

- Have we conducted a comprehensive data audit, or are blind spots holding us back?
- How much are data quality issues costing us in lost revenue or efficiency?
- Do we have clear governance structures to keep data accurate and trusted over time?

Building a Data-Driven Culture

While technology and tools get most of the attention in AI discussions, the truth is that culture often makes or breaks AI initiatives. You can invest millions in data platforms and machine learning models, but if your people don't value, trust, and use data, those investments won't pay off.

According to NewVantage Partners' (2022) annual survey of executives, 92% of firms cite cultural challenges, not technology, as the biggest barrier to becoming data-driven. This underscores a critical truth: preparing for AI is not just about cleaning databases; it's about shaping mindsets and behaviors across the organization.

From "Data as Exhaust" to "Data as a Strategic Asset"

In many organizations, employees see data as "digital exhaust." In other words, they see data as a byproduct of doing business, rather than a valuable business resource. For example, sales reps may skip entering full notes into CRM systems, or customer service agents may only partially log calls. These small omissions create massive blind spots when AI models attempt to learn from incomplete data. To build a data-driven culture, leaders must shift the narrative: data isn't just paperwork, it's the raw material for decision-making and innovation.

Education and Literacy

Employees at all levels need data literacy, which is the ability to understand, question, and use the data relevant to their roles. Data literacy isn't about turning everyone into data scientists; it's about helping staff know how their inputs affect AI outcomes.

For example:

- Marketing teams should understand how incomplete customer fields impact personalization accuracy.
- Operations managers should know why sensor calibration affects predictive maintenance models.
- Executives should be able to ask data-driven questions like, *"What assumptions underlie this forecast?"*

Organizations that invest in data literacy for their employees training gain greater trust in AI recommendations and faster adoption of analytics tools throughout the organization.

Accountability and Ownership

Data quality cannot be "IT's job" alone. Each department must take responsibility for the accuracy and completeness of its own data. This means:

- Defining data stewards in each team to oversee inputs.
- Creating accountability structures where errors are corrected, not ignored.
- Recognizing employees who model good data practices.

When data ownership is distributed across the business, the load is lighter and the outcomes stronger.

Leadership by Example

Finally, leaders set the tone. If executives make decisions based on intuition while ignoring data, employees will follow suit. If leaders ask for reports grounded in evidence, celebrate data-informed wins, and emphasize transparency, they reinforce the importance of data at every level. McKinsey research shows that organizations where senior leaders champion data use are 1.5 times more likely to report significant revenue growth from AI and analytics (McKinsey & Company, 2021).

Why Culture Is the Long-Term Differentiator

Technology will continue to evolve — today's cutting-edge platform may be tomorrow's legacy system. But a culture that values data is far more durable. It ensures that as tools change, the behaviors and disciplines that underpin AI readiness remain intact.

In short, a data-driven culture is the foundation that turns clean, integrated data into actionable intelligence. Without it, AI efforts risk becoming expensive experiments. With it, organizations create a self-reinforcing cycle: better data → better AI insights → better decisions → stronger belief in data.

- Do our employees treat data entry and accuracy as a priority — or as an afterthought?
- Are we investing in data literacy training to help staff understand how AI depends on their inputs?
- Do our leaders model data-driven decision-making consistently?

Ethical and Regulatory Dimensions of Data

If data is the fuel of AI, then ethics and regulation are the guardrails that keep it on track. Collecting, storing, and using data without clear safeguards is not just a technical risk — it's a reputational and legal liability. As businesses embrace AI, they must confront the dual challenge of using data responsibly while also navigating a complex and evolving regulatory landscape.

1. The Privacy Imperative: GDPR and CCPA

The European Union's **General Data Protection Regulation (GDPR)** and California's **Consumer Privacy Act (CCPA)** have set new global standards for data privacy. These laws require businesses to:

- Collect only the data that is necessary ("data minimization").
- Provide customers with visibility into what data is collected and how it is used.
- Allow individuals to access, correct, or delete their personal data.
- Protect data with robust security measures.

Non-compliance can be costly. GDPR fines can reach €20 million or 4% of annual global revenue, whichever is higher. Beyond fines, breaches of trust can drive customers away permanently. Cisco (2022) found that 48% of consumers have already switched companies over privacy concerns. For businesses preparing data for AI, compliance means more than checking boxes. It requires designing systems that balance personalization and predictive power with customer control and consent.

2. Bias in Training Data

AI models are only as fair as the data they are trained on. If historical data reflects human bias — in hiring decisions, lending approvals, or customer service interactions — AI can perpetuate or even amplify these inequities.

Examples include:

- Hiring algorithms that favored male applicants because historical hiring data was male-dominated.
- Loan approval systems that penalized minority applicants due to biased credit histories.
- Customer service bots that misunderstood or undervalued certain dialects and accents.

Mehrabi et al. (2021) warn that bias can enter at multiple stages — from data collection to labeling to algorithm design. Addressing this requires deliberate practices, including:

- Diverse training datasets that represent multiple groups fairly.
- Regular bias audits to test for disparate impacts.
- Explainability tools that help stakeholders understand how models make decisions.

3. Data Security and Stewardship

Data breaches remain one of the most pressing threats. A single breach can expose millions of customer records, triggering regulatory penalties and reputational damage. IBM's annual report (2023) estimated the average cost of a data breach at $4.45 million in 2023 — the highest on record.

For AI, security takes on new urgency because compromised data corrupts the entire decision-making pipeline. If attackers manipulate training data, they can cause AI systems to make flawed or even malicious recommendations. This makes data stewardship critical: encryption, access controls, continuous monitoring, and strong incident response protocols are not optional — they are prerequisites for trustworthy AI.

4. Transparency and Customer Trust

Ultimately, regulations like GDPR and CCPA reflect a deeper truth: customers want transparency. They want to know not only that their data is safe, but also that it is being used in ways that align with their interests and values. Businesses that communicate openly about their data practices — what is collected, why it's used, and how it benefits the customer — build stronger trust. In fact, customers are often willing to share more data if they perceive tangible value in return, such as personalized offers, improved service, or faster problem resolution.

Why Ethics and Regulation Matter for AI Readiness

Ethics and compliance are not hurdles to innovation. These principles are enablers of innovation. AI systems built on transparent, fair, and secure data foundations are more likely to be adopted, trusted, and scaled. Companies that ignore these principles may achieve short-term gains, but they risk long-term setbacks in the form of customer back-lash, lawsuits, or regulatory fines. In short, ethical and regulatory readiness is business readiness. Responsible data practices don't just protect the company; they create a competitive advantage by signaling reliability and trustworthiness in an AI-driven world.

📌 **Guiding Questions for Reflection**

- Do we have clear processes to comply with GDPR, CCPA, and other privacy laws?
- How are we identifying and addressing **bias in training data**?
- Are our security protocols strong enough to protect the integrity of our AI systems?
- Do our customers understand — and trust — how we use their data?

Case Studies: Data Readiness in Action

The importance of clean, structured, and well-governed data is not just theoretical. Leading organizations across industries have demonstrated that AI success depends on data readiness first. These cases illustrate how businesses that invested in data quality and integration reaped outsized benefits from AI.

Expanded Case Example: Starbucks — Personalization at Scale

The Challenge

Starbucks wanted to improve customer loyalty and revenue through personalized promotions. But with tens of millions of daily customers and fragmented data sources (loyalty cards, mobile app orders, POS systems), personalization was inconsistent and often generic.

The AI Solution

Starbucks centralized its customer data into a unified platform and invested heavily in cleaning and structuring that data. With a reliable foundation, it built "Deep Brew," its AI engine, which used machine learning to:

- Analyze individual purchase histories.
- Incorporate contextual data (time of day, weather, store location).
- Deliver personalized drink or food recommendations through the mobile app and loyalty program.

The Results

- Significant increase in revenue per customer through targeted offers.
- Loyalty program members became 2x more valuable than non-members.
- Starbucks could predict and suggest orders with a high degree of accuracy, strengthening customer engagement (Starbucks, 2019).

Lessons Learned

Data centralization and quality control transformed Starbucks' ability to personalize at scale. Clean, integrated data turned routine coffee purchases into curated experiences and showing that data readiness can directly boost loyalty and lifetime value.

Expanded Case Example: General Electric (GE) — Industrial IoT and Predictive Maintenance

The Challenge

GE operates in asset-heavy industries like aviation, energy, and manufacturing, where equipment downtime costs millions. GE had massive volumes of sensor data from turbines, jet engines, and industrial machines, but the data was noisy, inconsistent, and siloed across divisions.

The AI Solution

GE launched its Predix platform, focused first on cleaning and structuring IoT sensor data. The company:

- Developed robust data pipelines to standardize units, formats, and timestamp alignment across machines.
- Applied labeling and metadata to contextualize sensor readings (e.g., linking vibration signals to specific engine components).
- Integrated operational and maintenance records to enrich the predictive models.

The Results

- Improved predictive maintenance, allowing GE to forecast failures weeks before breakdowns occurred.
- Reduced downtime for industrial clients, saving millions in lost production.
- Opened new revenue streams by offering "as-a-service" predictive analytics to customers.

Lessons Learned

IoT sensors generate vast data, but without cleansing and structuring, it is little more than noise. GE's case shows that AI's value in industrial contexts comes only after mastering messy, high-volume data integration.

Expanded Case Example: Airbnb — Trust Through Data Integration

The Challenge

Airbnb faced a fundamental trust challenge: how to ensure hosts and guests felt safe engaging with strangers. While it had structured booking data, the platform also held massive amounts of unstructured content — reviews, messages, photos — that weren't being fully leveraged.

The AI Solution

Airbnb invested in cleaning, integrating, and labeling both structured and unstructured data. Its trust and safety AI systems now:

- Analyze text reviews and messages for fraud or safety signals.
- Assess photos of listings to verify authenticity and detect prohibited items.
- Integrate booking patterns with sentiment analysis to flag suspicious behavior.

The Results

- Significant reduction in fraudulent listings and bad actor activity.
- Stronger community trust, fueling continued growth of the platform.
- Faster dispute resolution between guests and hosts, backed by AI insights.

Lessons Learned

By combining structured and unstructured data, Airbnb created a **holistic trust model** that reassured both sides of its marketplace. The lesson: AI-enabled trust depends on **integrated, high-quality data across multiple sources**.

📌 **Guiding Questions for Reflection**

- Are we centralizing and cleaning our data the way Starbucks did before building personalization engines?
- Could predictive maintenance, like GEs, save us money by anticipating breakdowns before they occur?
- How could we combine structured and unstructured data, as Airbnb did, to strengthen trust and customer experience?

Closing Thought

AI may be the engine of business transformation, but data is the fuel that powers it. Without clean, structured, and well-governed data, even the most advanced AI tools will underperform. Worse, they can amplify errors, perpetuate bias, or erode customer trust. The promise of AI depends on what we feed into it — and that means making data readiness a top organizational priority.

This chapter has shown that data readiness is more than a technical exercise. It is:

- Strategic: Organizations that treat data as a core asset, like Starbucks, GE, and Airbnb, unlock new opportunities for growth, efficiency, and trust.
- Cultural: Building a data-driven culture ensures that employees across functions value and protect the integrity of data.
- Ethical: Transparent, compliant, and bias-aware practices not only protect against regulatory risk but also build stronger customer relationships.

The long-term winners in the AI economy will not just be those with the flashiest algorithms, but those with the cleanest pipelines, the strongest governance, and the deepest cultural commitment to data integrity.

For leaders, the takeaway is clear: before asking *"What can AI do for us?"*, ask *"Is our data ready for AI?"*. Because in the end, AI doesn't magically fix bad data — it magnifies it. Businesses that make the investment in data quality and governance today will be the ones that reap the rewards of AI tomorrow.

📌 Guiding Questions for Reflection

Preparing your business for AI isn't just about technology — it's about ensuring your data foundation is strong enough to support it. Use these questions as a self-assessment tool to evaluate your organization's readiness:

1. Visibility and Ownership

- Do we know where all our business-critical data resides?
- Who "owns" the accuracy, maintenance, and accessibility of our data?

2. Quality and Completeness

- What percentage of our data is "clean" (accurate, up to date, and consistent)?
- Are we losing opportunities or revenue due to duplicates, missing fields, or outdated records?

3. Integration and Accessibility

- Are our systems still siloed, or do we have a unified view of customers, products, and operations?
- Could APIs, data warehouses, or master data management help us consolidate?

4. Culture and Literacy

- Do employees across the business treat data as a strategic asset, or as a back-office task?
- Are we investing in data literacy to ensure staff understand how their inputs affect AI outcomes?

5. Ethics and Compliance

- Are we fully compliant with GDPR, CCPA, or other privacy regulations in the regions we serve?
- How are we identifying and addressing bias in our training data?
- Do our customers trust us to handle their data transparently and responsibly?

6. Future Readiness

- If we launched an AI initiative tomorrow, would our data be ready?
- What is the single biggest data obstacle holding back our AI ambitions today?

Please refer to the below appendix located at the end of the book for chapter-specific resources. However, be aware that inclusion in the appendices should not be considered an endorsement by the author for any individual commercial product.

APPENDIX 2: AI-Powered Data Capture & Utilization Platforms

CHAPTER 4
AI-DRIVEN DECISION MAKING

EXECUTIVES TODAY FACE AN OVERWHELMING PARADOX: while businesses have access to more data than ever before, decision-making often lags behind reality. According to IDC (2022), global data creation is projected to reach 181 zettabytes by 2025, yet only a fraction of this data is analyzed in time to inform decisions. Traditional tools focus on descriptive analytics, which is basically explaining what happened in the past. AI, by contrast, enables predictive and prescriptive analytics, which includes forecasting future outcomes and recommending actions. This evolution means organizations can move from reactive reporting to proactive, evidence-based decision-making to grow their business. But despite these unprecedented advances, the goal of AI cannot be to replace human judgment. Complex business decisions still require experience, context, and values. AI's role is to augment executive intuition with richer insights and real-time intelligence. Leaders who combine human expertise with AI-powered decision tools are better positioned to adapt in uncertain markets.

Data Visualization and Dashboards

For years, dashboards have been the go-to tool for business intelligence. Executives and managers alike have relied on colorful bar charts, pie charts, and line graphs to track performance. But here's the catch: while these visuals look neat, they often stop short of providing real insight. A pie chart can tell you what portion of sales came from a certain region last quarter, but it can't tell you why sales dipped in that region or whether it's likely to happen again next quarter.

This is where AI fundamentally changes the game. Instead of static snapshots, modern dashboards have become living, interactive decision aids. Platforms like Tableau and Microsoft Power BI don't just display data — they analyze it in real time, highlight patterns, and even surface hidden relationships you might never notice on your own.

Think about it this way: instead of you spending hours poking around the data to figure out what's unusual, the dashboard now proactively tells you, *"Hey, something's off here — you might want to look closer."* These systems can:

- Highlight anomalies automatically — maybe a sudden spike in expenses, or a sharp drop in customer retention.
- Detect correlations that aren't obvious — like how weather patterns might influence product sales, or how website traffic trends align with conversion rates.
- Provide forecasting widgets that help you look around the corner — anticipating not just what's happening now, but what's likely to happen next.

The shift is profound. Leaders are no longer just reviewing history; they're getting a forward-looking, narrative-rich picture of their business. As Stephen Few (2012) notes, the goal of visualization is not just to display numbers but to tell the story behind them. With AI-powered dashboards, the story is told faster, clearer, and with a lot more intelligence baked in.

For example, imagine sitting in a weekly leadership meeting. Instead of someone presenting last month's static sales numbers, an AI-driven dashboard might show, in real time, that revenue is trending lower in one product line. AI might then suggest that this dip can be correlated with a competitor's promotions, and even project how long the dip might last if nothing changes. That's the difference between reporting data and making decisions with data.

Ultimately, dashboards are evolving from rearview mirrors into navigation systems. They don't just help you see where you've been. With AI, they can now help you understand where you're going, and how to adjust your route along the way to manage unforeseen circumstances.

📌 Reflection Questions

- Do our current dashboards stop at "what happened," or do they also explain "why" and "what's next"?
- Are our teams still manually digging for insights that AI could surface automatically?
- How could interactive, AI-powered dashboards change the way we run our weekly or monthly business reviews?

Expanded Case Example: Tableau with AI-Powered Insights

The Challenge

Picture a regional sales manager getting ready for the monthly performance review. She opens the standard dashboard and sees what she always sees: sales by product line, sales by region, quarter-over-quarter growth. The numbers tell her something is off. They indicate revenue in the Southeast region dipped 12%. But they don't explain why. *Was it a product issue? A pricing change? A competitor entering the market?* Without a data science team on call, she's left with more questions than answers.

This scenario is all too common. Traditional dashboards are great for displaying "what happened," but they rarely go deeper. Leaders often need to rely on intuition or pull in analysts for weeks of extra digging just to understand the root cause of a trend.

The AI Solution

That's where Tableau's AI-powered features like "Explain Data" and "Ask Data" come in. Instead of staring at a static bar chart and guessing why revenue in the Southeast region dipped 12%, our sales manager can click directly on the anomaly and instantly get an explanation. The system automatically runs statistical tests across dozens of variables and returns likely drivers: maybe sales of a specific product line dropped, or perhaps competitor discounts surged in the same region. Even better, she doesn't need to learn SQL or be an analytics expert. She can simply type a natural language query into the dashboard, such as:

"Why did Southeast revenue drop last quarter?"

And within seconds, Tableau surfaces a set of clear, data-backed explanations, complete with visuals she can share with her team.

The Results

- What once took weeks of analyst time now happens in minutes.
- Business users without technical expertise can self-serve insights, reducing bottlenecks and empowering managers across the company.
- Adoption of analytics tools increases because they're finally intuitive and useful for day-to-day decision-making.
- Most importantly, leaders shift from reactive reporting to proactive action — making adjustments before issues spiral out of control.

Lessons Learned

Tableau's case shows how AI can turn dashboards from passive scorecards into active collaborators in the decision-making process. The key insight? When data becomes interactive, explanatory, and predictive, it doesn't just inform leaders — it empowers them to act with confidence.

📌 **Guiding Questions for Reflection**

- If your managers had access to AI-powered explanations like this, how much faster could they move from spotting problems to solving them?

Real-Time Decision Support

In today's business environment, waiting until the end of the month, or worse, the end of the quarter, to see performance results is like flying a plane with outdated maps. By the time you know there's a problem, it's often too late to fix it. This is where AI-driven, real-time decision support changes the game. Instead of looking backward at static reports, leaders can now access dashboards that are always up to date, pulling live data from sales systems, finance platforms, supply chains, and even external sources like market trends or weather patterns.

Think about what this means in practice. A CFO who once had to wait weeks for consolidated reports can now glance at a dashboard and instantly see the company's current cash position, updated down to the minute as invoices are paid and expenses are booked. If something unusual happens, say, a sudden spike in supplier costs or a dip in receivables, the system doesn't just bury it in a spreadsheet. It flags the anomaly immediately, often with AI-generated suggestions about what might be causing it.

Even better, these AI-powered tools don't just stop at showing the present. They look ahead. With built-in forecasting, executives can run "what-if" scenarios on the fly:

- *"If our marketing spend increases by 10%, how will that affect revenue in the next quarter?"*
- *"If supplier prices rise by 5%, how will margins shift?"*

This ability to test decisions before making them gives leaders a safety net that older systems simply couldn't provide. The impact is huge. Decision-making cycles get shorter. Leaders can act faster to capture opportunities or contain risks. And instead of being passive consumers of after-the-fact reports, executives and managers can become active navigators of the business, steering important decisions in real time with confidence.

In many ways, AI-driven decision support is like moving from a paper map to GPS navigation. The map still tells you where you are, but the

GPS updates constantly, reroutes around obstacles, and warns you about what's ahead. That's the difference between "reporting" and true decision support.

📌 Guiding Questions for Reflection

- Which of our decisions are still based on stale, after-the-fact reports?
- How would our leadership conversations change if we had access to live, predictive KPIs?
- What's one area of the business where real-time visibility could prevent surprises?

Expanded Case Example: CFOs Using AI Dashboards for Real-Time Key Performance Indicators (KPI)

The Challenge

Imagine being a CFO at a multinational company in today's volatile market. Every decision you make about spending, cash flow, or capital allocation can ripple across the organization. Yet, despite the stakes, many CFOs still rely on monthly or quarterly financial reports. That means by the time a red flag shows up, such as an unexpected rise in supplier costs or a sudden dip in revenue collections, weeks have already passed. The opportunity to intervene early is gone, and the team is left scrambling to contain the damage. As one CFO put it in a McKinsey survey: *"It often feels like we're driving by looking in the rearview mirror."*

The AI Solution

Now, imagine instead logging into a dashboard that updates itself constantly. The moment an invoice is paid, a sale is closed, or an unexpected expense comes through, the numbers shift instantly. That's what AI-powered financial dashboards deliver: real-time visibility into key metrics like cash flow, operating margin, and customer acquisition costs.

These systems don't just track numbers — they actively watch for unusual patterns. If expenses suddenly spike in a particular business unit, the dashboard doesn't just record it. The dashboard alerts the CFO immediately and even suggests potential causes. Many of these tools also include forecasting engines, so leaders can test scenarios on the spot:

- *"What happens to our liquidity if we speed up accounts receivable by five days?"*
- *"If fuel prices rise 10%, how does that affect our quarterly margins?"*

The CFO isn't just crunching numbers anymore. They're running live simulations of the company's financial future.

The Results

- Decision-making cycles shrank dramatically — from weeks to hours.
- Forecasting errors dropped, in some cases by nearly 40% (McKinsey, 2020).
- CFOs shifted from being rearview "scorekeepers" to strategic copilots, guiding the business with real-time intelligence.

Lessons Learned

The big takeaway is that AI doesn't just make finance faster; it makes it smarter. With always up-to-date insights, CFOs can act proactively instead of reactively, steering the organization through uncertainty with confidence. For many companies, this has meant catching risks before they snowball — and spotting opportunities they might otherwise have missed.

📌 **Guiding Questions for Reflection**

- What decisions in our organization would benefit from real-time visibility?
- Do our executives rely on outdated reports for critical decisions?
- How can AI dashboards improve responsiveness to financial risks and opportunities?

- If your finance leaders had access to live dashboards that not only showed today's numbers but also projected tomorrow's outcomes, how much more agile could your organization be?

Natural Language Querying of Data

One of the biggest challenges in building a data-driven culture is that most employees aren't data scientists. They don't know SQL, they don't want to wait two weeks for IT to run a custom report, and they often feel shut out of the analytics conversation. The result? Decision-making stays concentrated at the top or in specialized teams, while the rest of the organization runs on gut instinct.

Natural language querying is changing that. Imagine being able to "talk to your data" the same way you talk to a colleague. Instead of learning complicated syntax or navigating endless dropdown menus, you just type or even say:

- *"What were sales in the Midwest last quarter?"*
- *"Which product line had the highest growth in June?"*
- *"Show me customer churn rates over the past six months."*

Tools like *Tableau's Ask Data*, *Microsoft Power BI's Q&A*, and *ThoughtSpot* make this possible. They interpret your plain-language question, run the analysis in the background, and return a chart, graph, or table within seconds. The power here isn't just in the speed. The power's also in the accessibility! Suddenly, every manager, not just the analysts, can get answers on demand.

Case Example: Retail Manager with Ask Data

The Challenge

Consider a retail manager overseeing 50 stores. She's constantly asked questions like: *"Why are sales down in the Northeast this week?"* or *"Which products are selling fastest after the new promotion?"* In the past, she'd have to email the analytics team, wait days for a report, and hope it answered the right question. By the time she got the data, the promotion might already be over.

The AI Solution

With a tool like Tableau's Ask Data, she can simply type her question into the dashboard:

- *"Why did Northeast sales drop this week?"*

The system immediately scans the underlying data and generates an answer:

- A line chart showing sales decline.
- An annotation explaining that the dip is linked to inventory shortages in two high-volume stores.

Instead of digging through spreadsheets, she gets an instant, actionable explanation.

The Results

- Decisions that used to take days now happen in minutes.
- Store managers no longer feel dependent on IT or analysts — they can explore the data themselves.
- The company sees more widespread use of analytics, because the tools feel approachable, not intimidating.

Lessons Learned

Natural language querying democratizes analytics. By lowering the barrier to entry, it empowers managers and employees across all levels to make data-driven decisions. The lesson here is simple: when more people can "ask the data," organizations move faster, smarter, and with more confidence.

📌 **Guiding Questions for Reflection**

- How many people in our organization can directly query the data they need?
- Would natural language tools reduce bottlenecks in our reporting processes?

- How could democratizing data access shift our culture toward more informed, confident decision-making?

Augmenting—Not Replacing—Executive Intuition

One of the biggest fears executives express when talking about AI is: *"Is this going to replace me?"* The short answer is no. AI is incredibly powerful at spotting patterns, crunching numbers, and surfacing probabilities. But it doesn't understand context, values, or nuance in the way humans do. It can tell you what's likely to happen, but not whether that outcome aligns with your company's culture, ethics, or long-term vision. That's where executive intuition still matters — and always will. Think of AI as a co-pilot. The system handles calculations, monitors conditions, and suggests potential routes. But the pilot, in this case the organization's executive or manager, is still responsible for deciding whether to take off, change course, or land.

Blending Data with Judgment

Here's a real-world example: A consumer goods company uses AI to recommend price adjustments for its products. The model suggests raising prices on a popular household item because demand is inelastic. On paper, the numbers make sense. But the CEO knows that customers are highly sensitive to brand loyalty in this product category, and a price hike might erode trust long-term.

In this case, the AI is right from a short-term profitability standpoint, but the executive's judgment is what prevents a decision that could hurt the brand in the bigger picture. The power comes from combining both: AI analyzes the data, the executive or manager decides what to do with the analysis.

Trusting (and Questioning) the Machine

Building confidence in AI-driven recommendations requires transparency. When executives understand *why* an algorithm suggested a certain action, they're more likely to trust it, and just as importantly, they're more prepared to challenge it when needed. That's why many organizations are investing in explainable AI (XAI): systems that don't just deliver outputs, but show the reasoning behind them. Executives who learn to ask, *"What does the model see that I don't?"* can use AI to sharpen their intuition rather than second-guess it.

A Mindset Shift

In practice, this means leaders don't view AI as competition but as a partner. Instead of feeling threatened by automation, they lean on it for the heavy lifting — forecasting, anomaly detection, scenario modeling — while reserving their energy for strategy, negotiation, and vision.

AI can answer *"What's likely to happen?"*

Leaders still answer *"What should we do about it?"*

📌 **Guiding Questions for Reflection**

- Are we using AI to complement executive judgment, or trying to outsource decisions altogether?
- Do our leaders have enough visibility into *why* AI suggests certain actions to build trust in it?
- What types of decisions in our business are best left to human judgment, even with AI input?

Closing Thought

If there's one big takeaway from this chapter, it's this: AI is a partner in decision-making, not a substitute. AI excels at crunching massive datasets, spotting trends humans might miss, and running endless "what-if" scenarios at lightning speed. But it can't capture the human side of business. That involves judgment, ethics, empathy, and vision. Values AI can't compute. Leaders who thrive in the AI era will be those who know how to blend machine intelligence with human intuition. Over time, organizations that embrace this balance will see powerful benefits. Decisions will become faster, more accurate, and more forward-looking, supported by real-time insights instead of outdated reports. More employees will be empowered to access and understand data, creating a truly data-driven culture. And executives will be free to focus on what they do best: setting direction, weighing trade-offs, and building trust.

The long-term opportunity is bigger than efficiency. Cultivating AI-augmented management and leadership means creating organizations that are more resilient, more adaptable, and more innovative. In a world of constant disruption, those qualities may prove to be the most valuable assets of all.

CHAPTER 5
FINANCIAL
OPTIMIZATION WITH AI

IF STRATEGY IS the brain of a business, then money is its lifeblood. Cash pays employees, fuels marketing campaigns, sustains operations, and funds the innovations that shape future growth. When cash flows are strong, a business can move with confidence, pursue opportunities, and weather downturns. When cash flows stall, even the most brilliant strategy struggles to survive. This is why financial optimization isn't just about bookkeeping or hitting quarterly targets, it has to be about ensuring the organization has the financial resilience to grow, adapt, and thrive.

Why Financial Optimization Is Critical for Growth and Resilience

For decades, business leaders have known that "cash is king." Profits on paper mean little if a company cannot manage liquidity, anticipate costs, or allocate resources wisely and timely. Growth requires capital, and resilience requires agility in how that capital is deployed. The businesses that outperform competitors are often not those with the most ambitious ideas, but those with the sharpest financial discipline. Financial optimization provides the stability to fund innovation while also cushioning the shocks of economic cycles, supply chain disruptions, or sudden shifts in consumer demand.

Mid-career professionals, in particular, understand the delicate balance of risk and return. You've likely seen companies overspend during booms and then struggle when conditions tighten. Optimized financial management helps leaders avoid those extremes by building foresight into both investment and cost control.

The Limits of Traditional Financial Management

The challenge, of course, is that traditional financial management tools and processes often fall short. Spreadsheets remain the workhorse in many organizations, but they are static, error-prone, and time-consuming to update. Forecasts based purely on historical data provide a limited view. This approach assumes tomorrow will look much like yesterday. And in today's volatile business environment, that assumption rarely holds.

Manual processes also keep finance teams stuck in reactive mode. By the time numbers are reconciled, reports generated, and variances analyzed, the opportunity to act has often passed. Leaders are forced to make critical decisions with incomplete or outdated information, which undermines confidence and increases risk. For many businesses, "budget season" feels like an exercise in ritual rather than a tool for true strategy.

From Rearview Mirror Reporting to Forward-Looking Strategy

Artificial intelligence (AI) is transforming this picture. Instead of operating like a rearview mirror—telling you where you've been but not where you're going—AI turns financial management into a forward-looking system. With the ability to analyze massive amounts of structured and unstructured data at lightning speed, AI provides predictive insights, real-time scenario planning, and smart recommendations that allow you to optimize the use of financial resources. For example, AI-powered tools can detect early signs of cost inflation, forecast cash flow risks weeks in advance, and run "what-if" models that help executives test different strategies before committing resources. This means finance teams shift from being scorekeepers of the past to strategic advisors shaping the future.

In practical terms, this is the difference between hiking with a compass and paper map for direction, or driving with a GPS that not only shows where you are now, but predicts the traffic ahead, offers alternate routes, and recalculates continually as traffic and road conditions change. The destination may still be yours to choose, but the journey becomes faster, safer, and more efficient.

AI in Forecasting and Budgeting:

Forecasting and budgeting have always been the financial compass of a business. They tell leaders where they believe the organization is headed and how much fuel it will take to get there. The problem, of course, is that traditional methods often feel more like guesswork than guidance. Spreadsheets can only capture so many variables, and historical data does not always account for sudden shifts in markets, customer behavior, or global events. This is why many managers view budget season as frustrating: they spend significant time building a plan that may unravel within weeks or months.

AI offers a way to change this cycle. Rather than relying solely on static, backward-looking numbers, AI systems can integrate information from multiple sources—historical sales, seasonality, supply chain metrics, customer demand, and even macroeconomic indicators—to create forecasts that are more accurate, dynamic, and responsive.

- **Predictive Analytics for Revenue and Expense Projections**

Predictive analytics combines historical data with machine learning algorithms to identify patterns and anticipate future outcomes. For instance, an AI-driven tool might forecast that sales will rise in the third quarter not only because of seasonal trends but also because it detects growing consumer interest in related products on social media. Similarly, it could predict higher shipping costs based on oil futures data that no human analyst would have the time to process manually.

- **Real-Time Scenario Modeling**

A major challenge in budgeting is how quickly assumptions become outdated. AI allows leaders to run "what-if" scenarios instantly: *What if revenue drops by 15%? What if marketing spend increases by 10%? What if interest rates rise another half-point?* Instead of waiting days or weeks for analysts to build models, AI can produce those forecasts in real time, giving leaders agility to adjust strategy as conditions change.

- **Variance Analysis Made Smarter**

Most managers know the drill: set a budget, compare it to actuals, and spend meetings explaining why reality did not match expectations. AI makes variance analysis smarter by not only flagging discrepancies but also identifying probable causes. For example, if expenses exceed projections, AI might detect that a vendor's rates have been creeping upward or that overtime hours surged in one department. This shifts variance analysis from *problem spotting* to *problem solving*.

The Payoff: Confidence in Decisions

When forecasts are more accurate and budgets more adaptive, leaders gain confidence. Instead of second-guessing whether the numbers are reliable, they can focus on making decisions: where to invest, what to cut, and how to position for growth. For mid-career professionals, this means less time buried in spreadsheets and more time influencing strategy at the executive table.

Case in Point: Microsoft has adopted AI forecasting across its finance operations, reducing forecasting errors and accelerating quarterly reporting. Finance teams spend less time on manual data collection and more time analyzing implications for long-term business growth (Microsoft, 2021).

- Do my current forecasts feel reliable enough to make confident decisions—or are they just "best guesses"?
- How quickly can my organization adjust budgets when conditions change?
- What business scenarios do I wish I could test more easily— and how could AI help?
- Am I spending more time gathering numbers than actually using them to influence strategy?

AI in Fraud Detection and Risk Management:

Fraud is one of the oldest business risks, but in today's digital economy, fraud has taken on new forms and scales. Organizations today face numerous and aggressive financial threats, including unauthorized transactions, vendor fraud, cybersecurity breaches, and financial misreporting, all of which can erode profits, damage reputations, and undermine stakeholder trust. In fact, the Association of Certified Fraud Examiners (ACFE) estimates that businesses lose approximately 5% of their revenue to fraud each year, translating into trillions of dollars globally (ACFE, 2022). Traditional fraud detection methods, such as manual audits, rule-based monitoring, and retrospective reviews, often identify issues only after damage has occurred.

AI is transforming this landscape by enabling businesses to detect, prevent, and even predict fraudulent activity in real time. By analyzing massive datasets for anomalies, AI can identify and flag suspicious patterns far faster and more accurately than human reviewers or static rule-based systems.

- **AI for Anomaly Detection in Transactions**

Machine learning algorithms excel at spotting patterns in financial transactions. Unlike static rules (e.g., flagging only transactions above a certain amount), AI models learn from historical data to recognize unusual behaviors—such as sudden spikes in expense claims, duplicate invoices, or deviations from typical customer payment habits.

These systems become smarter over time, continuously refining their ability to differentiate between legitimate exceptions and genuine red flags.

For instance, banks and payment processors now use AI-powered fraud detection systems that can analyze thousands of transactions per second, immediately blocking suspicious activity. This not only protects customers but also prevents businesses from suffering financial and reputational losses.

- **Identifying High-Risk Customers, Vendors, and Partners**

Risk management extends beyond transactions. AI can also evaluate relationships across the business ecosystem, assessing the likelihood of risk from vendors, suppliers, or even employees. By analyzing data sources such as credit histories, regulatory filings, and historical behavior, AI can score counterparties for risk exposure.

This is particularly valuable in supply chains, where a financially unstable or non-compliant supplier could disrupt operations or expose the company to regulatory penalties. AI-driven risk scoring gives leaders the foresight to diversify or renegotiate relationships before crises occur.

- **Compliance and Regulatory Risk Management**

Beyond fraud, AI helps manage compliance risks by automatically monitoring financial transactions against regulatory requirements (e.g., anti-money laundering rules, GDPR, SOX). Natural language processing (NLP) can even review contracts for clauses that may create hidden liabilities, helping businesses reduce exposure before agreements are signed. In highly regulated industries such as finance and healthcare, this proactive approach can be the difference between a smooth audit and costly penalties.

Case Example: JPMorgan's COiN Platform

A leading case study in AI-driven risk management is JPMorgan Chase's Contract Intelligence (COiN) platform. Traditionally, lawyers and analysts would manually review 12,000 new commercial credit agreements each year, a process requiring 360,000 hours of work. With COiN, AI reviews these documents in seconds, identifying potential risks, anomalies, or compliance issues with far greater speed and accuracy (JPMorgan Chase, 2017). This not only reduces risk but also frees human experts to focus on strategic legal and financial decisions.

📌 Guiding Questions for Reflection

- How does my organization currently detect fraud or financial irregularities—and how reactive is that process?
- Could AI-driven anomaly detection save us time and reduce false positives compared to our current methods?
- Which vendors, partners, or customers pose the greatest financial or compliance risks to my business?
- What would it mean for my leadership role if compliance reviews shifted from a reactive burden to a proactive advantage?

AI in Expense Optimization and Cost Control

Every business leader knows the pressure of balancing growth with cost discipline. Spending is easy. Optimizing spending is much harder. In many organizations, costs creep in quietly—unmonitored subscriptions, inefficient vendor contracts, excessive overtime, or supply chain bottlenecks. These inefficiencies may seem minor in isolation but, over time, they erode profitability and weaken financial resilience.

Traditional cost-control measures typically rely on manual expense reports, periodic audits, or broad budget cuts. While these methods can highlight obvious issues, they often miss subtler patterns or emerging inefficiencies. Worse, blanket cost-cutting can stifle innovation or reduce employee morale. What leaders need is a smarter, more nuanced approach—one that pinpoints waste without undermining value creation.

Artificial intelligence (AI) is now enabling precisely this. By analyzing spending data across multiple systems, AI tools can uncover hidden cost drivers, suggest optimization strategies, and even predict where inefficiencies are most likely to arise in the future.

- **AI-Powered Categorization and Analysis**

One of AI's strengths is its ability to categorize and analyze vast amounts of data far beyond the capacity of manual review. For instance, machine learning algorithms can automatically classify expenses into granular categories, flagging anomalies such as duplicate invoices, inflated travel claims, or recurring subscriptions that deliver little value.

AI-driven expense platforms can also benchmark company spending against industry peers, showing where costs are above average and offering insights into potential savings. This transforms expense reports from static snapshots into dynamic tools for financial strategy.

- **Identifying Hidden Cost Drivers**

Sometimes, costs don't balloon because of fraud or negligence, but because inefficiencies are baked into operations. For example, supply chains may generate higher costs due to poorly timed ordering cycles, or manufacturing lines may suffer from unnecessary downtime. AI can detect these inefficiencies by analyzing data streams from procurement, production, and logistics. Predictive models can then suggest changes such as renegotiating vendor terms, consolidating orders, or rescheduling work shifts to reduce overtime.

- **Smart Contract Analysis and Vendor Management**

Contracts with vendors and suppliers are often complex, and manual reviews miss opportunities for renegotiation. AI systems that leverage natural language processing (NLP) can scan contracts to identify unfavorable terms, overlapping services, or missed opportunities for volume discounts. This is particularly powerful for mid-sized companies that lack large procurement teams. A mid-tier manufacturing firm, for instance, used AI-driven vendor analysis to uncover that it was paying different prices for the same raw material across regions. By standardizing contracts, the firm reduced costs by 12% in a single year (Accenture, 2021).

Balancing Cost Optimization with Growth

The goal of expense optimization is not to cut for the sake of cutting, but to align costs with strategy. AI enables leaders to make targeted adjustments that protect growth areas while trimming waste. Instead of across-the-board cuts, businesses can reinvest savings into innovation, marketing, or workforce development—ensuring that financial discipline strengthens, rather than hinders, long-term competitiveness.

📌 Guiding Questions for Reflection

- Where in my organization might "hidden costs" be silently eroding profits?
- Do I have real-time visibility into expenses, or am I relying on periodic, backward-looking reports?
- Which vendor contracts could benefit from AI-assisted review and renegotiation?
- How can I ensure cost-cutting measures support long-term strategy instead of undermining it?

AI in Cash Flow Management

Cash flow is the pulse of a business. Unlike revenue or profit, which can be manipulated by accounting entries, cash flow shows whether a company can meet its obligations today, tomorrow, and next quarter. Many businesses that appear profitable on paper collapse simply because they cannot pay their bills on time or fund day-to-day operations. According to U.S. Bank, 82% of small business failures are attributed to poor cash flow management or poor understanding of cash flow (CB Insights, 2021).

Traditional approaches to cash flow management often rely on spreadsheets, historical comparisons, or manual invoice tracking. These methods are limited: they are reactive, error-prone, and rarely capture the complex variables that influence liquidity. AI is changing this by providing predictive, real-time insights into cash inflows and outflows, allowing leaders to anticipate risks and act before they become crises.

- **Predicting Late Payments**

One of the most common cash flow challenges comes from delayed customer payments. AI-driven invoicing platforms can analyze historical payment behavior, credit scores, and even industry patterns to forecast which invoices are at risk of being paid late. This gives finance teams the opportunity to follow up proactively, adjust credit terms, or prepare alternative financing.

For example, AI can detect that a particular customer has a history of paying invoices 15 days late during the fourth quarter, allowing the business to plan accordingly. Over time, predictive models reduce surprises and improve the accuracy of working capital forecasts.

- **Working Capital Optimization**

AI doesn't just identify risks; it helps optimize the balance between receivables, payables, and inventory. By analyzing supplier payment histories, sales patterns, and inventory turnover rates, AI systems can recommend adjustments such as extending supplier payment terms without straining relationships, accelerating collections from reliable customers, or reducing excess stock. This kind of optimization ensures businesses don't leave money tied up unnecessarily, freeing cash that can be reinvested in growth initiatives. For mid-career leaders, this shifts cash management from a reactive chore to a strategic advantage.

- **Scenario Simulation for Liquidity Planning**

Another strength of AI lies in scenario simulation. Finance teams can instantly test the cash flow impact of different strategic decisions: *What if we offer a 5% early payment discount to customers? What if supplier costs rise by 7%? What if sales underperform by 20%?* These simulations allow leaders to make proactive decisions backed by data, rather than reacting once liquidity problems surface.

Case Example: Small Business AI Invoicing

A small e-commerce company adopted an AI-powered invoicing system that not only automated billing but also flagged high-risk invoices. Within six months, the company reduced late payments by 30%, stabilized its cash flow, and redirected saved time into growth activities. This demonstrates that AI-driven cash flow management is not just for large enterprises—it is increasingly accessible to small and mid-sized businesses as well.

📌 Guiding Questions for Reflection

- Do I know, with confidence, which customers are most likely to delay payments next quarter?
- How much cash is currently "trapped" in receivables or excess inventory?
- What scenarios could put my business at risk if cash flow tightened suddenly—and am I prepared for them?
- How could AI tools give me earlier warning signals than I currently have?

AI in Democratizing Finance

For years, advanced financial management capabilities were reserved for large enterprises with big budgets and dedicated finance teams. Smaller firms had to make do with spreadsheets, entry-level accounting systems, and a lot of manual oversight. This created a gap: the businesses that could afford sophisticated forecasting, risk analysis, and expense optimization often pulled ahead, while small and mid-sized businesses (SMBs) struggled to keep pace.

AI is helping to close this gap. Thanks to cloud-based platforms and affordable software-as-a-service (SaaS) solutions, advanced AI-powered finance tools are now accessible to companies of almost any size. This "democratization" of financial intelligence means that small businesses can apply the same principles of automation, forecasting, and optimization that once required enterprise-level resources.

- **Cloud-Based AI Platforms for SMBs**

Modern AI-powered finance applications—like QuickBooks, Xero, and Zoho Books—now come with built-in features such as automated expense classification, predictive cash flow forecasting, and anomaly detection. These tools were once the domain of Fortune 500 finance departments but are now available as part of low-cost monthly subscriptions. By leveraging these platforms, SMBs can gain real-time visibility into their finances, reduce manual bookkeeping errors, and make data-driven decisions without needing large finance teams.

- **AI Assistants for Finance Teams**

Even within larger organizations, AI is helping to "level the playing field" for finance professionals who are not data scientists. Natural language processing (NLP) tools enable managers to pose questions in plain English, such as "What has been our cash flow trend over the last six months?" They receive instant responses in the form of dashboards, graphs, or projections. This not only saves time but also empowers more leaders across the organization to participate in financial decision-making.

- **Leveling the Competitive Playing Field**

Democratizing finance with AI doesn't just save time and reduce errors, it gives smaller organizations a competitive advantage. A local retailer with AI-driven inventory and expense tracking can now compete with larger chains by making more efficient purchasing decisions. A mid-sized manufacturer using AI to optimize vendor contracts can improve margins almost as effectively as multinational corporations.

According to a Deloitte survey, 63% of SMBs using AI in finance reported higher profitability within two years, compared with only 37% of peers that relied on traditional methods (Deloitte, 2021). The takeaway is clear: AI doesn't just benefit the "big players" anymore. In many cases, it provides smaller firms with the agility to outmaneuver larger competitors.

Case Example: QuickBooks Cash Flow Forecasting

Intuit's QuickBooks AI now offers predictive cash flow management to its small business customers. The system can forecast up to 90 days ahead by analyzing past invoices, bills, and seasonal trends. Small businesses using the feature have reported fewer liquidity surprises and greater confidence in making short-term investments (Intuit, 2021). What used to require a finance department and advanced modeling tools is now available to a single business owner with a laptop and an internet connection.

📌 **Guiding Questions for Reflection**

- Am I fully using the AI-driven features already built into my accounting or ERP software?
- What financial processes still consume significant manual effort in my business—and could AI tools reduce some of that effort?
- How could democratized AI tools give my smaller business an edge over larger competitors?
- If my finance team had AI assistants answering routine questions, what higher-value work could they focus on?

The Chief Financial Officer's Role in an AI-Driven World

The role of the Chief Financial Officer (CFO) has evolved dramatically over the past two decades. Once considered the company's "chief accountant," focused on reporting and compliance, today's CFO is expected to be a strategic advisor who drives growth, manages risk, and shapes the long-term vision of the business. AI is accelerating this transformation, shifting the CFO's role from backward-looking score-keeper to forward-looking strategist.

According to a PwC global survey, 72% of CFOs believe that AI and automation will fundamentally reshape their function within five years (PwC, 2021). This doesn't mean AI will replace finance leaders—instead, it will empower them to influence broader business outcomes with sharper insights, faster analysis, and more confident decision-making.

From Number-Cruncher to Strategic Advisor

AI now handles many of the time-consuming tasks that once consumed finance teams: reconciliations, variance analysis, expense categorization, and compliance checks. With those processes automated, CFOs are free to focus on higher-level strategic questions: *How do we deploy capital for maximum impact? Where should we invest in innovation? How do we balance short-term performance with long-term resilience?* This shift is placing CFOs increasingly at the center of digital transformation and positioning them not just as financial stewards and strategists, but as enterprise-wide change agents.

CFOs must take on the role of trust-builders. The effectiveness of AI systems depends on the quality of the data they use and the models they are based on. Leaders must validate, audit, and clarify AI-generated forecasts for boards, investors, and employees. Transparency and explainability are essential. If executives don't understand how AI arrived at its conclusions, they're less likely to act on them.

This means CFOs must become bilingual in both finance and technology. They must be able to interpret AI outputs while ensuring compliance, fairness, and alignment with organizational goals.

Upskilling Finance Teams for an AI Future

The future of finance is not just about adopting AI tools but about cultivating teams that know how to use them effectively. CFOs must lead reskilling initiatives, ensuring that finance staff can partner with AI rather than feeling threatened by it. This may involve training in data literacy, scenario modeling, and strategic analysis.

By positioning AI as a partner rather than a competitor, CFOs can foster a culture where human expertise and machine intelligence complement each other. This "augmented intelligence" model allows finance professionals to spend less time preparing data and more time applying judgment.

Case Example: CFOs Driving Transformation

Global consumer goods companies such as Unilever have demonstrated how CFOs can use AI to influence enterprise strategy. By adopting AI-powered forecasting and scenario modeling, Unilever's finance leaders improved supply chain agility while guiding investment in sustainable products. Instead of just reporting results, the CFO's office became a driver of business transformation (Unilever, 2021).

📌 **Guiding Questions for Reflection**

- In my role (or my organization's finance leadership), how much time is spent on manual reporting versus strategic decision-making?
- Do I trust AI-generated financial insights enough to present them confidently to boards or stakeholders?
- What skills will my finance team need in the next 3–5 years to thrive in an AI-driven environment?
- How can I position finance not as a back-office function, but as a strategic partner driving organizational growth?

Closing Thought

AI doesn't replace the need for financial acumen in a business organization; it simply redefines the role. From forecasting to fraud prevention, expense control to cash flow, and democratization of tools to CFO leadership, AI gives businesses increased financial agility from the ability to optimize finances like never before. In a competitive, uncertain world, financial optimization and agility is the true foundation for survival.

Please refer to the below appendices located at the end of the book for chapter-specific resources. However, be aware that inclusion in the appendices should not be considered an endorsement by the author for any individual commercial product.

APPENDIX A: AI-Powered Forecasting and Budgeting Platforms

APPENDIX B: AI-Powered Fraud Detection & Risk Management Platforms

APPENDIX C: AI-Powered Expense Optimization and Cost Control Platforms

APPENDIX D: AI-Powered Cash Flow Management Platforms

APPENDIX E: AI-Driven Comprehensive Financial Management Platforms

APPENDIX F: Upskilling & Retraining Resources for Finance Teams Embracing AI

CHAPTER 6
WORKFORCE MANAGEMENT AND PRODUCTIVITY

EVEN IN AN ERA of digital transformation, people remain the most critical asset of any organization. Technology, capital, and strategy set the stage, but it is employees who ultimately execute plans, engage with customers, and drive innovation. Workforce management is therefore central to organizational performance. Businesses with effective workforce strategies outperform peers in profitability, customer satisfaction, and long-term resilience.

Yet today's workforce environment is more complex than ever. Hybrid work arrangements, global talent shortages, rising labor costs, and shifting employee expectations present leaders with unprecedented challenges. Traditional human resource practices, such as manual resume screening, periodic performance reviews, and standardized training, struggle to keep pace. They are often slow, biased, and reactive, leading to inefficiencies and disengagement.

AI is transforming workforce management by introducing greater precision, efficiency, and personalization. Rather than replacing human judgment, AI enhances it. By automating repetitive tasks, analyzing extensive employee and performance data, and generating predictive insights, AI allows managers to concentrate on leadership, strategy, and workplace culture. Over the past few years, there has been a surge in organizations that have implemented AI platforms for various aspects of workforce management, including recruitment, scheduling, and learning.

The potential benefits are clear: better hires, more engaged employees, optimized schedules, and targeted reskilling programs. But the stakes are also high. Poorly designed AI systems risk introducing new forms of bias, reducing transparency, or alienating employees who fear being managed by algorithms. Leaders must therefore balance efficiency with empathy, and automation with accountability.

This chapter explores how AI is being applied to recruitment and talent acquisition, employee engagement, productivity optimization, and continuous learning—and how leaders can adopt these tools responsibly to create workplaces that are both more efficient and more human.

🎯 Guiding Questions for Reflection

- How has workforce management changed in my organization over the past five years?
- What workforce challenges (e.g., hiring delays, turnover, engagement) consume the most time and resources today?
- Where could AI provide insights or automation that free leaders to focus on strategy and culture?
- What safeguards should I consider to ensure fairness and trust in AI-driven workforce decisions?

AI in Recruitment and Talent Acquisition

Recruitment has always been one of the most critical yet resource-intensive aspects of workforce management. A bad hire can cost a company between 50% and 200% of the employee's annual salary when factoring in recruitment costs, training, lost productivity, and disruption to teams. In competitive labor markets, organizations can't afford inefficiencies or biases in their hiring processes.

Traditional recruitment often relies on manual resume reviews, subjective interview impressions, and slow scheduling processes. These methods not only delay hiring but also risk overlooking qualified candidates due to human biases or time constraints. AI is transforming this landscape by making talent acquisition faster, smarter, and more inclusive.

- **Resume Screening and Candidate Filtering**

AI-powered recruitment platforms can efficiently analyze thousands of applications to identify candidates who are best suited for job openings. Instead of relying solely on keyword-based filters, which can overlook important context, modern machine learning systems assess entire resumes, cover letters, and even online profiles to evaluate qualifications, skills, and career paths This approach saves recruiters significant time, enabling them to concentrate on high-potential applicants.

- **Skill and Culture Matching**

Beyond basic screening, AI can evaluate whether a candidate's skills align with the competencies most predictive of success in a role. Some platforms use predictive analytics to assess a candidate's likelihood of excelling based on past performance data from current employees. Advanced systems even incorporate cultural fit assessments, analyzing values alignment and communication style through natural language processing.

- **Reducing Bias in Hiring**

One of the most promising applications of AI in recruitment is its potential to mitigate unconscious bias. Properly designed algorithms can focus on skills and performance indicators rather than demographic factors like gender, race, or age. For example, anonymized resume screening systems remove identifying information, helping companies widen their talent pools. However, a word of caution here is essential, because if historical hiring data contains bias, AI systems can inadvertently reproduce those patterns. This is a good example of why oversight and auditing of AI platforms is essential.

- **Candidate Experience and Efficiency**

AI doesn't just help recruiters. AI can also improve the experience for potential candidates. Chatbots and virtual assistants answer questions, schedule interviews, and provide updates in real time, creating smoother interactions and reducing candidate drop-off rates. Automated scheduling systems eliminate back-and-forth emails, enabling interviews to be arranged in minutes instead of days.

Case Example: Unilever's AI-Powered Recruitment

Unilever, one of the world's largest consumer goods companies, adopted an AI-driven recruitment process combining online games, AI-analyzed video interviews, and predictive algorithms. The system reduced hiring time by 75% and increased candidate diversity, while maintaining or improving the quality of hires (Harver, 2021). By automating repetitive tasks, recruiters were able to focus more on relationship-building and strategic workforce planning.

📌 **Guiding Questions for Reflection**

- Where in my organization's recruitment process do delays, inefficiencies, or biases most often appear?
- Could AI resume-screening or candidate-matching tools improve both speed and fairness?
- How might chatbots or virtual assistants enhance the candidate experience in my company?
- What safeguards should I put in place to ensure AI does not reinforce historical biases in hiring.

AI In Employee Engagement and Retention

An organization can have the best strategy and technology, but without engaged employees, execution falters. Employee engagement, or the emotional commitment employees have to their organization's goals, directly influences productivity, innovation, and retention. Gallup's most recent *State of the Global Workplace* report found that only 32% of U.S. employees are engaged at work, while 17% are actively disengaged, costing businesses billions annually in lost productivity (Gallup, 2022).

Turnover is another pressing issue. Replacing an employee can cost upwards of between 50% and 200% of their annual salary, depending on the role. High turnover not only drains organizational finances but also disrupts teams, erodes customer relationships, and stalls strategic initiatives. For mid-career managers and executives, finding sustainable ways to engage employees and retain top talent is therefore a core leadership priority.

AI is emerging as a powerful ally in addressing these challenges. By analyzing employee data, predicting risks, and personalizing engagement strategies, AI provides leaders with the insights needed to foster healthier, more productive workforces.

- **Sentiment Analysis for Real-Time Engagement**

Traditional employee engagement surveys are static and often outdated by the time results reach leadership. AI-enabled sentiment analysis tools can monitor emails, chat platforms, and employee surveys to capture real-time mood and engagement levels. Natural language processing (NLP) algorithms detect patterns in tone, word choice, and frequency of communication, providing HR leaders with dashboards that highlight emerging issues before they escalate. This shift allows leaders to address engagement challenges proactively rather than waiting for quarterly or annual surveys.

- **Turnover Prediction**

Machine learning models are increasingly used to predict which employees are at risk of leaving. By analyzing factors such as tenure, promotion history, workload, manager interactions, and even external labor market trends, AI can assign "attrition risk scores" to employees. Armed with this information, HR can then intervene with tailored retention strategies such as career development conversations, workload adjustments, or financial incentives. This predictive capability reduces surprise resignations and stabilizes workforce planning.

- **Personalized Retention Strategies**

AI also supports the personalization of employee engagement. There are AI platforms available that can recommend learning opportunities, internal mobility paths, or well-being programs that align with each employee's skills, interests, and aspirations. Personalized development and recognition increase employee satisfaction and loyalty, especially among younger generations who value career growth and individualized support.

Case Example: IBM's Predictive HR Analytics

IBM has been a pioneer in applying AI to workforce management. Its HR division uses predictive analytics to identify employees at risk of leaving with up to 95% accuracy. Managers receive targeted recommendations on interventions, ranging from mentorship to role redesign, which has saved the company millions annually in turnover costs (IBM, 2019). By integrating AI insights into HR workflows, IBM has transformed retention from a reactive problem to a proactive strategy.

📌 **Guiding Questions for Reflection**

- How does my organization currently measure engagement, and how quickly can we act on the results?
- Could predictive analytics help identify employees at risk of leaving before they resign?
- What personalized development or recognition strategies could AI suggest for my team?
- How do I balance AI-driven insights with human judgment to maintain trust and empathy?

AI in Workforce Productivity and Task Automation

Workforce productivity has always been a central concern for business leaders. Productivity is not just about how hard employees work, it's about how effectively they spend their time and energy. Studies show that employees lose an average of 20–30% of their working hours to repetitive, low-value tasks such as data entry, scheduling, or searching for information (McKinsey Global Institute, 2018). These tasks, while necessary, often drain morale and limit time for creative, strategic, or customer-facing activities that drive real value.

AI and automation technologies are reshaping productivity by taking over routine, repetitive work and optimizing resource allocation. Instead of replacing employees, AI enables them to focus on higher-value tasks. The shift is away from doing more work to doing better work, i.e. work smarter, not harder.

- **Robotic Process Automation (RPA) for Repetitive Tasks**

Robotic Process Automation (RPA) involves software "bots" that replicate rule-based human actions in systems such as finance, HR, or customer service. These bots can process invoices, reconcile data, and update records at speeds far beyond human capacity. When integrated with AI, RPA becomes "intelligent automation," able to make decisions based on context rather than just rules. For finance teams, this could mean automated expense approvals. For HR teams, it might mean faster onboarding document processing. In both cases, automation reduces errors, cuts costs, and frees employees for more meaningful work.

- **Intelligent Scheduling and Shift Optimization**

Scheduling is another area where inefficiency undermines productivity. Manual scheduling often leads to misaligned staffing, excessive overtime, or employee burnout. AI-enabled workforce management systems can forecast demand, predict peak workloads, and assign shifts accordingly. For example, retail stores can use AI to predict foot traffic and adjust staffing, while hospitals can balance nurse schedules with patient volume projections. Optimizing employee scheduling creates not only cost savings but also improves employee satisfaction by aligning workloads with capacity.

- **AI-Enhanced Collaboration and Knowledge Management**

Collaboration platforms like Microsoft Teams, Slack, and Zoom are increasingly embedding AI to improve workflow. These systems can:

- Suggest meeting times and agenda items.
- Summarize conversations automatically.
- Recommend relevant documents or colleagues to consult.

Such tools reduce "digital friction" — the time wasted navigating multiple apps or repeating administrative tasks — and instead channel attention toward decision-making and problem-solving (Gartner, 2022).

Expanded Case Example: AI Scheduling in Retail

The Challenge

A national retail chain with over 500 stores faced ongoing difficulties in managing its workforce. Store managers struggled to balance staffing levels with fluctuating customer traffic. During peak shopping times, long checkout lines frustrated customers, while during slower periods, stores were overstaffed, driving up labor costs.

Traditional scheduling relied on manager intuition and static sales forecasts, which failed to account for real-time factors such as local events, promotions, weather patterns, or even regional shopping behaviors. This reactive approach contributed to employee dissatisfaction (due to unpredictable shifts and overtime) and lost revenue (from poor customer service at peak hours).

The AI-Driven Solution

The company implemented an AI-powered workforce management system to optimize staffing across all locations using:

- Data Integration: The system aggregated data from point-of-sale transactions, loyalty card usage, weather forecasts, and even local event calendars.

- Predictive Analytics: AI models predicted hourly customer traffic for each store, identifying likely peak and slow periods.
- Optimized Scheduling: Using these forecasts, the platform automatically generated staffing rosters that aligned employees' skills (e.g., cashiers vs. floor staff) with anticipated demand.
- Employee Flexibility: Staff could access schedules via a mobile app, request shift swaps, and receive instant notifications of changes.

The Results

- Cost Reduction: Labor costs decreased by 15% due to reduced overstaffing in slow periods.
- Customer Satisfaction: Net Promoter Scores (NPS) improved by 12% as checkout times shortened and floor coverage improved during peak hours.
- Employee Morale: Worker surveys showed increased satisfaction thanks to more predictable schedules and easier shift swapping.
- Managerial Efficiency: Store managers spent 50% less time on scheduling, redirecting energy toward sales initiatives and employee coaching.

Lessons Learned

1. Human Oversight Matters: While AI produced highly accurate schedules, local managers retained flexibility to make adjustments for personal employee needs.
2. Transparency Builds Trust: By explaining how demand forecasts worked, the company reduced skepticism among employees who initially resisted "algorithmic scheduling."
3. AI as a Strategic Enabler: Scheduling wasn't just about cost-cutting; it became a strategic advantage by aligning staffing with customer experience goals.

This retail case demonstrates how AI-driven workforce productivity tools can simultaneously lower costs, improve employee satisfaction, and enhance customer experience. Rather than replacing human decision-makers, the system provided data-driven guidance, allowing managers to make more confident, fair, and strategic choices.

📌 Guiding Questions for Reflection

- Which tasks in my organization consume significant time but contribute little strategic value?
- Could AI or RPA tools automate these processes without reducing quality or compliance?
- How might intelligent scheduling improve both efficiency and employee satisfaction?
- Are my employees empowered with AI-enhanced collaboration tools, or are they still stuck in manual workflows?

AI in Reskilling and Continuous Learning

Artificial intelligence and automation are transforming job roles across various industries. According to McKinsey, by 2030, there may be as 375 million workers globally who need to switch occupational categories or acquire new skills to remain employable (Manyika et al., 2017). This isn't limited to frontline workers. AI will eventually impact all areas of the workforce. Fields like finance, HR, IT, medicine, even fields considered artistic, such as music composition or creative writing, are all seeing parts of their roles impacted by AI. This impact is creating demand for higher-level analytical, strategic, and human-centric skills. Reskilling is therefore no longer a "perk" or a side initiative. Reskilling an organizations workforce is a strategic necessity for business resilience. Organizations that invest in continuous learning gain a double advantage: they adapt more quickly to technological disruption and build stronger employee loyalty by demonstrating commitment to career growth.

- **AI-Powered Skill Gap Analysis**

Traditional skills assessments often rely on manager evaluations or employee self-assessments, which can be biased or incomplete. AI-driven platforms, however, analyze various workforce data, including job descriptions, performance metrics, and even trends in the external labor market. This analysis helps map existing employee skills against the competencies that companies will need in the future.

This gap analysis allows leaders to identify which roles are most at risk of disruption and where targeted reskilling investments will yield the greatest return. Research from MIT Sloan and others highlight trends occurring like "skills inference," which means using AI to analyze employee data to understand proficiency and guide workforce strategy. Major companies like Johnson & Johnson and DHL have begun using AI to assess workforce skills and plan training or internal hiring. However, experts recommend maintaining human oversight in the process. Gap Analysis matters because The World Economic Forum estimated 44% of worker skills will be disrupted in some way by the year 2025 (World Economic Forum, 2020). Employers are increasingly turning to AI to help identify gaps and reskill their workforces.

- **Personalized and Adaptive Learning Platforms**

Modern AI learning systems provide personalized and adaptive training. By analyzing an employee's career goals, learning history, and skill gaps, these systems recommend tailored training modules, which can range from data analysis to leadership development. Additionally, some platforms utilize "microlearning," offering short and focused lessons that can be accessed as needed. This approach helps employees integrate training into their busy schedules more easily. For instance, AI-powered platforms like Coursera for Business and Degreed use recommendation algorithms similar to those of Netflix or Spotify. They suggest courses that are most relevant to each employee's role and aspirations.

- **Microlearning and Just-in-Time Training**

Instead of long, infrequent training sessions, AI supports on-demand learning. Employees can receive prompts when new skills are needed. For instance, a project manager might get a 10-minute module on predictive analytics before leading a data-driven project. This kind of just-in-time training ensures knowledge is immediately relevant and retained.

Case Example One: AT&T's $1 Billion Reskilling Initiative

The Challenge

By the mid-2010s, AT&T faced a stark reality: its business was rapidly shifting from traditional telecommunications to a digital-first ecosystem centered on cloud computing, cybersecurity, and advanced networking. An internal audit revealed that nearly half of its 250,000 employees lacked the skills necessary for the company's future direction (AT&T, 2018).

This left AT&T at a crossroads. One option was to aggressively recruit external talent, competing in an already tight labor market. The other was to invest in its existing workforce. Leadership realized that the latter was both more cost-effective and more sustainable, particularly in maintaining employee loyalty and organizational culture.

The AI-Driven Solution

AT&T invested $1 billion over several years to launch one of the largest corporate reskilling initiatives ever undertaken. Central to the program was an AI-powered career platform that:

- Assessed employee skills using performance data, resumes, and self-reported capabilities.
- Mapped skill gaps by comparing current employee competencies with the future roles AT&T anticipated needing in cybersecurity, software engineering, and data science.

- Recommended training paths customized to each employee's starting point and career goals, drawing from partnerships with Coursera, Udacity, and Georgia Tech's online degree programs.
- Gamified progress tracking, providing feedback loops and incentives to keep employees motivated.

This system allowed employees to see transparently which roles were most at risk of obsolescence and what steps they could take to remain competitive within the company.

The Results

- Massive Participation: More than 100,000 employees engaged in the program, pursuing everything from short courses to full master's degrees.
- Talent Redeployment: Thousands transitioned into high-demand areas such as cybersecurity, data analytics, and software engineering, filling roles that would otherwise have required costly external hires.
- Retention Boost: Employees reported greater loyalty to the company, as AT&T demonstrated a willingness to invest in their long-term careers.
- Cultural Transformation: The initiative helped shift AT&T's culture toward continuous learning, making upskilling an expected part of career progression.

Lessons Learned

1. Investing in People Pays Dividends: Retaining and reskilling existing employees proved more cost-effective than large-scale external hiring.
2. Transparency Builds Trust: Employees appreciated seeing clear pathways from their current roles to future opportunities.
3. Partnerships Expand Capacity: Collaborating with universities and online learning providers ensured high-quality, scalable training resources.

4. AI Personalization is Key: By tailoring learning paths to individuals, the program avoided a one-size-fits-all approach and kept engagement high.

Why It Matters

The AT&T initiative has since become a benchmark case study in corporate reskilling. It demonstrated that with sufficient investment and AI-enabled personalization, even massive workforces can be retrained for the digital economy. More importantly, it highlighted the strategic value of continuous learning not just as a workforce necessity but as a competitive advantage in fast-moving industries.

Case Example Two: Siemens and Manufacturing Reskilling

The Challenge

Siemens, a global leader in manufacturing and industrial automation, faced a workforce transformation challenge as advanced robotics, IoT (Internet of Things), and AI began reshaping factory operations. Many employees were skilled in traditional mechanical engineering but lacked digital and data competencies. This created a gap between current capabilities and the skills needed to operate "smart factories."

The AI-Driven Solution

Siemens developed an AI-driven skills assessment and training ecosystem. The platform:

- Analyzed employee job roles and performance data to identify digital skill gaps.
- Recommended personalized learning paths in areas like data analytics, IoT integration, and predictive maintenance.
- Delivered training through microlearning modules that workers could access during downtime on factory floors.

The Results

- Upskilling at Scale: Thousands of employees transitioned into hybrid roles combining mechanical expertise with digital proficiency.
- Increased Productivity: Plants with reskilled workers reported efficiency gains of up to 20%, thanks to better integration of AI-enabled machines.
- Employee Retention: Workers reported higher morale, viewing Siemens as an employer invested in their future rather than one replacing them with machines.

Lessons Learned

- Practicality Matters: Delivering training in short, on-demand modules fit the realities of factory work.
- AI as a Guide: Employees trusted AI recommendations when paired with human mentoring and coaching.

Case Example Three: Walmart and Retail Reskilling

The Challenge

As e-commerce expanded, Walmart needed to reskill its workforce for roles in digital fulfillment, supply chain automation, and customer service analytics. Traditional retail skills — stocking, cashiering — were increasingly supplemented (or replaced) by tasks requiring technological literacy.

The AI-Driven Solution

Walmart launched its LiveBetterU program, enhanced by AI-based personalization engines:

- Employees received individualized career pathways using AI that matched their current roles, skill sets, and long-term aspirations.

- The system suggested relevant courses (e.g., supply chain logistics, data analysis, or customer engagement) delivered through online platforms like Guild Education.
- AI also predicted which employees were most likely to succeed in advanced training, ensuring resources were allocated effectively.

Results

- Large-Scale Impact: More than 400,000 employees participated in the program by 2022, with thousands transitioning into tech-enabled retail and logistics roles.
- Improved Retention: Walmart reported higher retention among employees who engaged in training compared to peers.
- Future-Proofing: By reskilling frontline workers, Walmart reduced its reliance on external hires for emerging roles.

Lessons Learned

- Accessibility is Key: Walmart subsidized tuition and integrated learning into employees' schedules, making participation feasible.
- AI + Career Mobility: Personalized recommendations gave employees confidence that reskilling would open real career opportunities, not dead ends.

📌 **Reflections from Industry Cases**

These case studies from manufacturing (Siemens) and retail (Walmart) illustrate that AI-driven reskilling is not limited to tech companies:

- In manufacturing, AI-enabled reskilling bridges the gap between mechanical expertise and digital factory operations.
- In retail, AI-powered personalization allows frontline employees to transition into higher-value roles in logistics, analytics, and customer engagement.

Both highlight a universal lesson: AI doesn't just transform jobs — it transforms learning itself, making reskilling more targeted, scalable, and impactful across industries.

Balancing Technology and Human Development

While AI enhances reskilling, leaders must ensure these programs remain human-centric. Employees may feel anxious about automation or skeptical of algorithmic career recommendations. Transparency in how skill assessments are made, combined with coaching and mentoring, ensures employees see AI not as a threat but as a partner in career growth.

📌 Guiding Questions for Reflection

- What roles in my organization are most at risk of automation or disruption in the next 3–5 years?
- Do we have real visibility into our workforce's current skills and future needs?
- How could AI-powered platforms personalize training for different career paths?
- Are we pairing AI-driven insights with human mentorship and support?

Balancing Efficiency with Human-Centric Management

AI is redefining how organizations hire, train, and deploy their workforce. From predictive scheduling to engagement analytics, AI promises unmatched efficiency. However, focusing too narrowly on cost reduction or productivity metrics risks eroding the very trust and morale that drive long-term performance.

A Gartner (2022) survey found that while 70% of organizations implementing AI workforce tools reported efficiency gains, nearly 40% faced pushback from employees citing concerns about fairness, transparency, and a loss of human connection. In short, AI can improve processes, but if employees feel "managed by algorithms," engagement may decline (Zhang et al., 2021).

- **The Need for Human-Centric Design**

The most effective organizations approach AI in workforce management as a supportive partner rather than a replacement for human leadership. This requires:

- Transparency: Employees should understand how AI makes decisions (e.g., scheduling or performance analytics).
- Fairness: Algorithms must be tested and monitored for bias, especially in hiring and promotions.
- Empathy: Managers must interpret AI insights through a human lens, recognizing that people's lives can't be reduced to data points.

In practice, this means AI suggests — humans decide. AI may recommend optimal schedules, but managers adjust based on personal circumstances. AI may flag disengagement risks, but leaders still hold one-on-one conversations.

- **Augmented Management: A Blended Approach**

Scholars increasingly advocate for "augmented management," where AI handles repetitive analysis and provides insights, while humans exercise judgment, empathy, and contextual decision-making (Shrestha et al., 2019). This blended approach ensures that technology amplifies human leadership rather than undermines it. For example, at Unilever, AI-driven employee sentiment analysis highlights engagement trends, but HR leaders still lead workshops and design interventions rooted in culture and values. At Deloitte, AI tools predict attrition risks, but managers use these predictions as starting points for coaching conversations, not as deterministic verdicts.

- **Ethical Considerations in Workforce AI**

Beyond operational efficiency, workforce AI raises ethical considerations:

- Privacy: How much employee data should organizations monitor?
- Consent: Are employees informed when AI is used to assess engagement or productivity?
- Bias: Are algorithms perpetuating historical inequities in hiring or promotion?

Organizations that fail to address these types of questions risk reputational damage, regulatory penalties, or internal distrust. Conversely, those who embrace ethical AI practices strengthen both compliance and employee loyalty.

Case Example: Amazon and the Risks of Over-Automation

One of the largest retailer's warehouse practices are often cited as a cautionary tale of AI-driven workforce management. The company employs algorithms to monitor worker productivity and even trigger automated termination when performance drops below certain thresholds. While this system improved short-term efficiency, it also led to reports of burnout, high turnover, and worker dissatisfaction (Palmer, 2021).

The lesson: automation without empathy may yield short-term efficiency gains but at the cost of long-term sustainability. Human oversight and ethical safeguards are essential to prevent workforce management from devolving into purely transactional relationships.

🎯 **Guiding Questions for Reflection**

- How do I ensure AI supports — rather than replaces — human judgment in workforce management?
- What safeguards are in place to ensure fairness, transparency, and trust in AI-driven decisions?

- Am I balancing efficiency goals with the human need for autonomy, recognition, and purpose?
- Could over-automation create risks of burnout or disengagement in my workforce?

Closing Thought

Organizational efficiency is crucial for survival, but efficiency should never come at the expense of the human workforce. The future of workforce management depends on achieving a balance: utilizing AI for automation and optimization while ensuring that leaders provide context, empathy, and ethical guidance. Organizations that find this balance will not only experience increased productivity but will also create workplaces where employees feel valued, trusted, and empowered. The competitive advantage will come not from AI alone, but from the collaboration between AI and human leadership.

Please refer to the below appendices located at the end of the book for chapter-specific resources. However, be aware that inclusion in the appendices should not be considered an endorsement by the author for any individual commercial product.

APPENDIX G: AI-Powered Recruitment & Screening Platforms

APPENDIX H: AI-Powered Employee Engagement & Retention Platforms

APPENDIX I: AI-Powered Workforce Productivity & Task Automation Platforms

APPENDIX J: AI-Powered Skills-Gap Analysis Platforms

CHAPTER 7
MARKETING AND CUSTOMER ENGAGEMENT WITH AI

MARKETING HAS ALWAYS BEEN about connecting the right message with the right customer at the right time. For decades, businesses relied on broad demographic categories, intuition, and retrospective analysis to guide marketing campaigns. While effective to a degree, this approach often resulted in wasted spend, generic messaging, and missed opportunities. Today, however, AI is transforming marketing into something far more dynamic, precise, and customer-centric.

At its core, AI allows marketers to turn data overload into actionable insight. Modern consumers generate vast amounts of behavioral data, from online shopping habits and search queries to social media engagement and even voice interactions with smart devices. A single customer journey may span dozens of touchpoints across different platforms, leaving traditional analytics tools overwhelmed. AI, by contrast, excels at analyzing massive, complex datasets in real time, uncovering patterns that humans would struggle to detect. The result is a fundamental shift: marketing is moving from reactive to predictive. Instead of analyzing what customers did last quarter, AI can anticipate what they are likely to do tomorrow. For example, predictive models can identify which products a shopper is most likely to buy next, which customers are most at risk of churn, or what time of day an email campaign will achieve the highest engagement.

Equally transformative is AI's ability to power personalization at scale. Consumers increasingly expect tailored experiences, such as relevant recommendations, individualized sales promotions, and seamless interactions across channels. A survey by McKinsey (2021) found that 71% of consumers expect companies to deliver personalized interactions, and 76% get frustrated when this doesn't happen. AI makes it possible to deliver this personalization to millions of customers simultaneously, something human marketers simply cannot achieve alone. But AI's influence extends beyond just targeting and personalization. It is also changing the tools of engagement. Conversational chatbots now provide 24/7 support, generative AI produces marketing copy and product descriptions in seconds, and dynamic pricing engines adjust offers in real time based on customer demand and competitor actions. These capabilities are not only driving efficiency but also raising the bar for customer expectations across industries.

However, this transformation is not without risks. Over-reliance on AI can lead to ethical concerns, such as privacy intrusion or misuse of personal data. Then there's the "creepiness factor" of overly intrusive personalization. The challenge for leaders is to leverage AI in ways that enhance trust rather than undermine it. This is a balancing act that requires ongoing scrutiny. In this new era, AI is not merely another tool in the marketer's toolkit. AI represents a paradigm shift. Just as the printing press, radio, and the internet redefined customer engagement in their eras, AI is reshaping how businesses connect with consumers in the 21st century. So, the question is not whether to adopt AI in marketing, but how to do so responsibly, effectively, and strategically.

🎯 Guiding Questions for Reflection

- How well does my current marketing strategy adapt to rapidly changing customer expectations?
- Am I using customer data primarily for retrospective reporting, or am I leveraging predictive insights?

- Where could personalization improve both customer trust and conversion rates in my business?
- How do I balance AI-powered efficiency with ethical concerns like privacy and transparency?

Customer Data Insights

In the digital age, customer data is both abundant and essential. Every search query, product review, click, and purchase adds to the massive flow of information that defines modern consumer behavior. Yet this abundance of data can easily overwhelm organizations that lack the tools to make sense of it. A survey by Forbes Insights (2019) found that 95% of businesses struggle with managing unstructured data, while only 29% reported being able to use their data effectively. AI offers a way forward by transforming raw information into actionable insights.

- ### Raw Data Becomes Actionable Intelligence

Traditional data analytics tools rely heavily on human-designed queries and static reporting. They can answer questions such as, *"What were last quarter's sales?"* but often fall short of answering *"Why did sales change?"* or *"What will customers want next?"* AI-driven analytics fills this gap by applying machine learning (ML) and natural language processing (NLP) to uncover hidden patterns and predict customer needs. For example, clustering algorithms can segment customers into micro-groups based on behavior, while predictive models forecast who is most likely to make a purchase or defect to a competitor. This transformation allows businesses to shift from reactive reporting to predictive and prescriptive intelligence — moving from looking in the rearview mirror to anticipating what lies ahead.

- ### Predictive Analytics Anticipates Needs

One of AI's most powerful contributions to customer insights is its predictive capability. By analyzing past transactions, browsing behavior, and contextual data (such as location or device type), AI can forecast future actions. Retailers, for example, can identify customers who are likely to repurchase within the next month or who may be at risk of churn. Financial institutions can use AI to detect early signs of credit risk or cross-sell opportunities. Such insights not only improve targeting but also enhance customer experiences by ensuring that marketing messages are timely and relevant.

- **Balancing Data Power with Privacy Concerns**

The rise of AI-driven customer insights brings significant ethical considerations. Consumers are increasingly aware — and wary — of how their data is collected and used. Laws like the European Union's General Data Protection Regulation (GDPR) and California's Consumer Privacy Act (CCPA) enforce stricter requirements for transparency, consent, and security. Businesses must balance personalization with privacy, ensuring that insights are used to serve customers rather than exploit them. Failing to strike this balance risks damaging trust, which is a key driver of long-term brand loyalty.

Case Example: Netflix's Recommendation System

Netflix offers a well-known example of turning data into actionable insights. The platform collects billions of viewing events daily, including when viewers pause, rewind, or abandon shows. AI algorithms analyze this behavior to personalize recommendations for each user's home screen. The results speak volumes: Netflix reports that over 80% of content watched is driven by recommendations, significantly improving customer satisfaction and reducing churn (Netflix, 2020). This case underscores the potential of AI not only to interpret customer data but also to translate it into highly personalized engagement strategies.

📌 **Guiding Questions for Reflection**

- What types of customer data does my organization collect today, and how effectively do we use it?
- Am I primarily focused on descriptive analytics (what happened) or predictive analytics (what will happen)?

- How can AI help uncover patterns that human analysts might miss?
- What safeguards are in place to ensure ethical data use and compliance with privacy regulations?

Personalized Marketing at Scale

Personalization is no longer a luxury in marketing. It is an expectation. Modern consumers want brands to understand them, anticipate their needs, and deliver relevant messages across every channel. A survey by McKinsey found that 71% of consumers expect companies to deliver personalized interactions, and 76% become frustrated when personalization is lacking. At the same time, companies that excel at personalization generate 40% more revenue from these activities compared to their peers (McKinsey, 2021). The challenge for businesses is scale. It is one thing for a small shopkeeper to know their customers by name and tailor recommendations accordingly. It is another for global enterprises serving millions of customers across multiple markets. This is where AI delivers transformational value.

- **Hyper-Segmentation and Dynamic Targeting**

Traditional marketing segments customers into broad categories such as age, income, or geography. While useful, these categories overlook the nuances of individual behavior. AI enables hyper-segmentation, identifying thousands of micro-segments based on purchase history, browsing behavior, device use, and even real-time contextual data.

For example, rather than targeting "women aged 25–40," an AI system might create a micro-segment of "women aged 30–35, who browse eco-friendly products on mobile devices and make repeat purchases every three months." Campaigns tailored to these segments achieve significantly higher engagement rates than generic messaging.

- **Real-Time Personalization Across Channels**

AI also makes personalization dynamic and real-time. In e-commerce, recommendation engines adjust instantly to customer behavior, offering complementary products or discounts based on what a shopper places in their cart. In digital advertising, programmatic ad platforms use AI to decide in milliseconds which ad to serve to a user based on browsing history, location, and inferred interests. This responsiveness creates seamless, omnichannel experiences. A customer who browses a product on their phone might receive a follow-up email later that day, or see a personalized discount banner when revisiting the website. AI ensures consistency across platforms without requiring manual coordination.

- **Generative AI for Content Personalization**

Generative AI is pushing personalization even further by creating customized content at scale. Platforms like Jasper AI and Persado use natural language generation (NLG) to craft personalized email subject lines, ad copy, and landing pages that resonate with different customer personas. This allows marketers to test multiple variations rapidly, optimizing for tone, emotion, and conversion rates. Instead of sending a single promotional email to all subscribers, companies can send thousands of tailored versions, each reflecting the recipient's preferences and behavior.

Case Example: Amazon's Recommendation Engine

Amazon's recommendation system is one of the most sophisticated applications of personalization at scale. It uses collaborative filtering and deep learning algorithms to analyze billions of customer interactions and generate product suggestions tailored to each user.

Amazon reports that its recommendation engine drives 35% of total sales, underscoring how AI-powered personalization is not just a customer experience enhancer but also a significant revenue generator. The system adapts in real time — if a customer buys a baby stroller, the platform immediately suggests related items such as car seats, diapers, or toys (McKinsey, 2021). This level of personalization demonstrates how AI can replicate the attentiveness of a personal shopper but on a massive, global scale.

- How personalized are my customer interactions today, and at what scale?
- Could AI-powered hyper-segmentation reveal overlooked customer micro-groups?
- Am I leveraging real-time personalization, or do my campaigns still operate on static segments?
- How might generative AI expand my ability to deliver individualized content at scale?
- Where is the balance between helpful personalization and "creepy" overreach in my industry?

Conversational AI and Customer Support

Customer support is often where loyalty is won or lost. A study by Microsoft (2020) revealed that 90% of consumers consider customer service quality when choosing whether to do business with a brand, and 58% will switch companies after a poor service experience. Yet providing timely, personalized, and efficient support is increasingly challenging as customer expectations rise and inquiries multiply across channels. AI is reshaping customer service by enabling conversational interfaces, using chatbots, voice assistants, and automated messaging systems that provide real-time engagement, handle high volumes, and reduce wait times. These tools not only cut costs but also enhance the customer experience when deployed thoughtfully.

- **The Rise of AI-Powered Chatbots**

Chatbots powered by natural language processing (NLP) can interpret customer queries, provide relevant answers, and escalate complex issues to human agents when necessary. Unlike early scripted bots, today's AI assistants learn from interactions, improving accuracy over time. For businesses, this means scaling support without proportionally increasing headcount. For customers, it means 24/7 access to assistance without waiting in call queues. Some analysts predict that by 2027, chatbots will become the primary customer service channel for roughly 25% of organizations, reflecting their growing ubiquity.

- **Voice Assistants and Multimodal Support**

Conversational AI is not limited to text. Voice-based systems like Alexa, Google Assistant, and Siri are increasingly integrated into customer service. Banks, airlines, and retailers now allow customers to check account balances, track deliveries, or reschedule flights via voice commands. Multimodal support allowing customers to seamlessly switch between voice, chat, and human service ensures flexibility and convenience. This ability is critical in building customer trust and reducing friction in problem resolution.

- **Augmenting, Not Replacing, Human Agents**

Although AI is capable of managing routine inquiries, complex or emotionally sensitive issues still require human empathy. The most effective systems use a hybrid model, where AI triages requests and provides first-line responses, while human agents handle escalated cases. AI can also empower human agents by offering real-time assistance, such as suggesting knowledge base articles, recommending responses, or analyzing customer sentiment during calls. This reduces resolution times and improves consistency.

Expanded Case Example: Sephora's Virtual Assistant

The Challenge

Sephora, one of the world's largest cosmetics retailers, faced a challenge common to many global brands: how to provide personalized beauty advice to millions of customers across digital and physical touchpoints. Beauty shopping is highly individual, with customers seeking product recommendations based on skin type, preferences, or style. Providing this level of personalization at scale would have been nearly impossible through human staff alone.

The AI Solution

Sephora implemented conversational AI across multiple channels, including Facebook Messenger, Kik (a messaging platform popular at the time), its website, and mobile app. The chatbot integrated product databases, customer purchase histories, and AI-powered recommendation engines to provide:

- Personalized product recommendations based on customer queries or browsing history.
- Beauty tips and tutorials, offering step-by-step advice on application and product combinations.
- In-store booking functionality, allowing customers to schedule makeovers or consultations through chat.
- Integration with augmented reality (AR), so users could virtually "try on" products before purchase.

The Results

- Customers reported higher satisfaction due to instant, personalized support.
- The chatbot drove measurable increases in conversion rates, as product recommendations aligned closely with customer needs.
- Sephora reduced pressure on human agents by automating routine queries, freeing staff for higher-value consultations.
- The blend of AI + AR created a uniquely engaging customer journey, reinforcing Sephora's position as a digital leader in retail.

Lessons Learned

Sephora's case illustrates that conversational AI is not just about reducing service costs. AI can directly drive sales growth and brand loyalty when designed around customer needs. Sephora's integration of AI into both digital and physical shopping also showed the potential of conversational AI as an omnichannel tool.

Expanded Case Example: Bank of America's "Erica"

The Challenge

Bank of America serves over 67 million customers across the U.S. Traditional customer service channels, such as call centers and in-branch visits, were costly to scale and often left customers frustrated with long wait times. As digital banking adoption grew, the bank needed a way to provide personalized, real-time financial support to millions of customers without overburdening its human staff.

The AI Solution

In 2018, Bank of America launched Erica, an AI-powered virtual financial assistant embedded in its mobile banking app. Erica combines natural language processing, predictive analytics, and financial algorithms to support customers by:

- Answering everyday banking queries (e.g., account balances, transaction history).
- Providing proactive financial insights such as detecting recurring subscriptions or flagging unusually high charges.
- Assisting with bill pay, transfers, and credit card management.
- Offering personalized financial advice, such as suggesting ways to improve credit scores or reduce fees.

Erica continuously learns from billions of customer interactions, improving both accuracy and personalization over time.

The Results

- By 2022, Erica had handled over 1 billion interactions, making it one of the most widely used AI assistants in the financial sector (Bank of America, 2022).
- The bank reported increased customer satisfaction scores, as Erica reduced wait times and provided instant answers.
- Human call center staff were able to focus on more complex financial issues, improving service quality.

- Erica helped Bank of America lower operational costs while enhancing customer trust through proactive insights.

Lessons Learned

Bank of America's Erica demonstrates how conversational AI can move beyond "reactive support" to become a proactive advisor. The key lesson is that when AI systems provide value-added insights, not just basic answers, they can deepen customer engagement and foster long-term loyalty.

These case studies highlight two critical lessons:

1. Conversational AI can enhance both support and sales — Sephora showed how chatbots drive conversions by combining personalization with convenience.
2. AI assistants can evolve into trusted advisors — Bank of America's Erica demonstrates that conversational AI can provide proactive value, not just transactional service.

📌 **Guiding Questions for Reflection**

- How responsive is my customer support system today?
- Could AI-powered chat or voice assistants reduce wait times and improve customer satisfaction?
- Are my human agents empowered by AI insights, or are they still navigating issues manually?
- How do I balance efficiency with the need for empathy in customer interactions?

AI in Content Creation and Campaign Management

Content is the fuel of modern marketing. From social media posts and blog articles to ad copy and email campaigns, organizations rely on a steady stream of fresh, relevant, and compelling content to capture customer attention. Yet producing this content at the required volume, speed, and personalization has long been one of marketing's greatest bottlenecks. AI is transforming how content is both created and managed. By automating repetitive writing tasks, optimizing campaigns in real time, and enabling data-driven creativity, AI allows marketers to operate at a scale and efficiency that would have been impossible just a few years ago.

- **Generative AI for Content Production**

One of the most visible applications of AI in marketing is generative AI. These are tools that use natural language generation (NLG) and large language models (LLMs) to produce human-like text. Platforms such as Jasper, Copy.ai, and OpenAI's GPT-based models can generate ad copy, product descriptions, blog posts, and even video scripts within seconds. This is not about replacing human creativity but augmenting it. Generative AI can handle the repetitive or time-consuming aspects of content production, freeing human marketers to focus on strategy, brand storytelling, and creativity. For example, an e-commerce team can use AI to instantly create hundreds of personalized product descriptions optimized for SEO, while reserving human input for crafting brand voice and campaign themes.

- **Campaign Optimization with Predictive Analytics**

AI doesn't just help create content — it ensures that content reaches the right audience, at the right time, in the right format. Predictive analytics tools analyze historical campaign data, customer behavior, and contextual signals (such as device type or location) to recommend the best channels, timing, and messaging for campaigns. For instance, an AI platform might determine that a particular customer segment is more likely to engage with a push notification on a mobile device during evening hours, while another segment responds better to email in the morning. Marketers can then dynamically adjust campaigns based on these insights, improving return on investment (ROI).

- **Automated A/B Testing and Continuous Optimization**

Traditional A/B testing, which entails sending two versions of an ad or email to see which performs better, is slow and limited. AI automates this process by testing multiple variations simultaneously, analyzing performance in real time, and automatically shifting spending toward the best-performing content. This capability turns campaign management into a continuous learning loop, where every customer interaction feeds into the next optimization cycle. As a result, campaigns become more agile, adaptive, and effective.

Expanded Case Example: The Washington Post's "Heliograf"

The Challenge

The Washington Post, one of the largest newspapers in the U.S., needed a way to cover the massive volume of data-driven stories during high-profile events such as the 2016 U.S. presidential election and the 2016 Summer Olympics. Reporters could not realistically provide real-time updates on every election result, local contest, or sports statistic while also producing in-depth reporting. The challenge was how to expand coverage without expanding headcount or sacrificing quality.

The AI Solution

To address this, the Post developed Heliograf, an AI-powered natural language generation (NLG) system designed to automatically generate short, factual news updates.

- Election Coverage: During the 2016 election, Heliograf produced thousands of updates, including live vote tallies and localized election alerts.
- Sports Results: At the Olympics, it generated automated medal counts and event summaries.
- Integration with Human Reporting: Heliograf was not intended to replace journalists, but to complement them. While AI generated structured updates, human reporters provided investigative depth, context, and analysis.

The Results

- The system generated over 850 stories in its first year, many of which reached niche audiences underserved by traditional reporting (Montal & Reich, 2017).
- Heliograf freed human reporters to focus on higher-value journalism, reducing pressure to chase routine updates.
- It demonstrated that AI could increase coverage breadth without sacrificing credibility or journalistic standards.

Lessons Learned

The Washington Post's use of Heliograf illustrates how AI can scale content creation in contexts where timeliness and factual accuracy matter more than narrative depth. For marketers, the lesson is clear: AI-generated content can handle repetitive or data-heavy tasks, while humans should focus on creativity, storytelling, and brand voice.

Expanded Case Example: Coca-Cola's AI-Driven Marketing

The Challenge

As a global brand, Coca-Cola must balance consistent brand identity with localized personalization across dozens of markets. At the same time, consumer expectations for interactivity and creativity have risen dramatically. The company needed a way to not only produce content faster, but also to engage customers in co-creating the brand experience.

The AI Solution

In 2023, Coca-Cola partnered with OpenAI and Bain & Company to launch the "Create Real Magic" campaign. This AI platform that allowed consumers to design custom Coca-Cola-themed digital art using generative AI tools. Customers could merge iconic brand assets (like the Coca-Cola script or polar bear mascot) with their own creative input. The best user-generated designs were showcased on Coca-Cola's billboards in Times Square and London's Piccadilly Circus.

In parallel, Coca-Cola used machine learning for real-time social media sentiment analysis, allowing marketing teams to rapidly adapt messaging across campaigns. AI tools helped localize ads, test creative variations, and predict which designs would resonate in specific regions.

The Results

- The campaign created a viral wave of customer engagement, with thousands of user-generated artworks shared on social media.
- Coca-Cola reported stronger brand connection among Gen Z and millennial audiences, who value interactive, co-creative experiences (Forbes, 2023).
- By integrating AI into both creative production and campaign management, Coca-Cola accelerated content delivery while maintaining global brand consistency.

Lessons Learned

Coca-Cola's case shows how AI can elevate marketing beyond efficiency into interactive, participatory experiences. Instead of simply delivering personalized ads, brands can empower customers to become co-creators, deepening emotional connections. It also demonstrates the power of combining generative AI (content creation) with predictive AI (campaign management) for maximum impact.

Together, these cases highlight the dual role of AI in content and campaign management:

- For The Washington Post, AI expanded content coverage and relieved human workload in structured reporting.
- For Coca-Cola, AI unlocked creativity and engagement by making customers part of the brand narrative.

📌 Guiding Questions for Reflection

- Which parts of my content pipeline (e.g., product descriptions, ad copy, blog posts) could be streamlined with AI tools?
- Am I using AI primarily for efficiency, or also to unlock new creative possibilities?
- How can predictive analytics improve my campaign targeting and timing?
- What safeguards are needed to ensure AI-generated content aligns with brand values and avoids bias or misinformation?

Closing thoughts

The central theme is that AI significantly enhances human capabilities in various ways. It allows organizations to scale routine tasks efficiently, which frees up valuable time for employees to focus on higher-level functions such as strategic planning, storytelling, and building trust with clients and stakeholders. Through AI, businesses can also offer personalized experiences to customers, tailoring their services and communications in ways that resonate more with individual preferences and needs.

However, it's essential to recognize the disparity in resources between large corporations and smaller businesses. Industry leaders like Bank of America and Coca-Cola possess substantial financial and technical resources, enabling them to develop bespoke AI platforms that cater specifically to their operational requirements and strategic goals. In contrast, smaller organizations may lack the same level of investment but can still tap into a range of commercially available AI solutions. While these "off-the-shelf" tools might not offer the same degree of customization as those developed in-house, they can still deliver effective results. By leveraging these accessible resources, smaller businesses can gain a competitive edge, improve their operations, and ultimately enhance their customer engagement and satisfaction.

Please refer to the below appendices located at the end of the book for chapter-specific resources. However, be aware that inclusion in the appendices should not be considered an endorsement by the author for any individual commercial product.

APPENDIX K: AI-Powered Customer Data Platforms (CDPs)

APPENDIX L: AI-Powered Personalized Marketing Platforms

APPENDIX M: AI-Powered Customer Service Platforms

APPENDIX N: AI-Powered Platforms which offer both Campaign Management and Predictive Analysis

CHAPTER 8
AI-POWERED PRICING AND SALES OPTIMIZATION

PRICING IS one of the most powerful levers for business growth, yet it is often one of the least optimized. Research has shown that a 1% improvement in price realization can translate into an 8–12% increase in operating profits, far outpacing the impact of equivalent improvements in sales volume or cost reduction (Marn & Rosiello, 1992). Despite this outsized influence, many organizations continue to rely on outdated methods such as static markups, competitor benchmarking, or seasonal discounting. These approaches, while simple to execute, frequently leave money on the table and fail to respond to rapidly changing market dynamics.

The digital economy has amplified the challenges of traditional pricing. Customers today can instantly compare prices across competitors, making transparency and agility essential. At the same time, the explosion of transaction data, customer behavior signals, and external factors (such as weather, macroeconomic conditions, and social media trends) creates an opportunity to fine-tune prices with unprecedented precision. But the sheer volume, velocity, and variety of data involved is well beyond human capacity to manage effectively.

This is where AI enters the picture. AI enables businesses to treat pricing not as a static decision, but as a dynamic and adaptive system. Machine learning models can analyze billions of data points in real time to forecast demand, anticipate customer willingness to pay, and recommend optimal prices across products, markets, and customer segments. For example:

- An airline dynamically adjusts ticket prices by route, seat class, and time until departure.
- A retailer sets localized prices based on regional demand and competitor actions.
- A Software as a Service (SaaS) platform, such as subscription Anti-virus software, uses predictive models to personalize subscription bundles for different customer profiles.

AI also brings a strategic shift from reactive adjustments to proactive optimization. Instead of waiting for end-of-quarter results to see how pricing performed, managers can use AI dashboards to monitor in-the-moment price elasticity, test multiple price points, and identify opportunities to improve margins.

Yet the promise of AI-powered pricing is not without risks. Overly aggressive algorithms can lead to "sticker shock," eroding trust if customers feel exploited, as in the case of surge pricing controversies in ride-hailing platforms. Moreover, ethical and regulatory concerns around algorithmic transparency and potential price discrimination are growing, particularly in markets governed by the EU's General Data Protection Regulation (GDPR) and emerging U.S. consumer protection guidelines.

In this context, pricing is no longer just a tactical decision but a strategic capability. Companies that can harness AI for intelligent pricing will gain a durable competitive advantage. However, they must balance profitability with fairness and transparency. For business managers the challenge is twofold: to understand how AI transforms pricing mechanics, and to navigate the organizational, ethical, and cultural shifts that come with it.

📌 Guiding Questions for Reflection

- How is pricing currently determined in my organization — intuition, competitor benchmarking, or data-driven analysis?
- Are we leaving revenue on the table by relying on static markups or one-size-fits-all discounting?
- What types of data (customer behavior, competitor activity, market signals) could AI leverage to improve pricing decisions?
- How can we ensure that AI-driven pricing balances profitability with fairness and customer trust?

Dynamic Pricing with AI

Pricing has always been about balancing supply and demand, but traditional methods often rely on broad seasonal patterns, manual markups, or competitor tracking. These approaches fail to capture the nuances of real-time market conditions. Dynamic pricing powered by AI can fundamentally change this equation by adjusting prices in real time based on shifts in demand, customer behavior, competition, and contextual factors such as time of day, weather, or location. AI-driven dynamic pricing systems use machine learning models to analyze massive datasets and predict customer willingness to pay. They enable businesses to set prices that optimize revenue, improve inventory utilization, and respond instantly to market fluctuations. According to McKinsey (2018), companies that adopt dynamic pricing strategies can improve margins by 2–5% on average, a significant impact in competitive industries.

How AI Powers Dynamic Pricing

- Demand forecasting: AI models predict demand for specific products or services based on historical sales, seasonality, and real-time signals.
- Elasticity modeling: Machine learning identifies how sensitive customers are to price changes in different contexts.
- Competitive intelligence: Algorithms scrape competitor prices and adjust accordingly.

- Contextual factors: AI incorporates weather, local events, or even traffic conditions into pricing models (useful in travel, hospitality, and ride-sharing).

These systems are self-learning, meaning they continuously refine their predictions as new data becomes available. However, commercially available AI-powered pricing platforms are often industry-specific, because pricing challenges differ dramatically by industry.

Expanded Case Example: Airlines and Hospitality

The Challenge

Airlines and hotels face a fundamental challenge: their inventory is perishable. A seat on a plane or a night in a hotel room loses all value once the flight departs or the night passes. Historically, pricing was managed manually, with seasonal adjustments, advance-purchase discounts, and blanket promotions. These methods were blunt instruments, leaving revenue on the table in high-demand scenarios and over-discounting in low-demand periods.

The AI Solution

Airlines such as Delta and Lufthansa, and hotel groups like Marriott International, turned to AI-powered revenue management systems. These systems integrate machine learning algorithms that analyze:

- Booking trends across hundreds of routes and properties.
- Customer booking windows (e.g., last-minute travelers vs. early planners).
- Competitor pricing and capacity.
- External signals such as weather forecasts, holidays, or conferences.

By continuously adjusting prices in real time, AI maximizes revenue per available seat or room. Marriott's system, for example, recalibrates rates daily to optimize both occupancy and profitability.

The Results

- Airlines implementing AI-driven pricing improved overall yield management, often outperforming manual systems by 3–5% (Cross et al., 2009).
- Marriott reported measurable gains in revenue per available room (RevPAR), particularly in competitive urban markets.
- Customers benefited from more consistent pricing availability, with fewer "over-discounting" errors that traditionally undermined profitability.

Lessons Learned

Dynamic pricing in airlines and hospitality illustrates that AI does not arbitrarily inflate prices. Instead, AI matches value with demand. When communicated clearly (e.g., advance-purchase savings vs. last-minute surcharges), dynamic pricing improves both revenue and customer satisfaction.

Expanded Case Example: Uber and Ride-Hailing

The Challenge

Uber faced the challenge of balancing driver supply with rider demand in real time. During peak events, such as concerts, storms, or rush hour, demand would spike, leaving riders stranded and drivers overwhelmed. Traditional flat pricing models could not solve this mismatch, leading to frustrated customers and lost revenue opportunities.

The AI Solution

Uber introduced its now-famous surge pricing model, powered by AI algorithms that monitor:

- Active rider requests in a given area.
- Available driver supply.
- Traffic conditions and travel times.
- Historical demand patterns.

When demand outpaces supply, the algorithm increases fares dynamically. The higher prices incentivize drivers to head toward high-demand zones while rationing demand among riders willing to pay more.

The Results

- Surge pricing improved service reliability, ensuring riders could still find a car during peak times.
- Drivers benefited from higher earnings in high-demand windows.
- However, Uber also faced public backlash, with riders accusing the company of price gouging, particularly during emergencies.

To address this, Uber adjusted the system to display clear fare multipliers and provide upfront fare estimates before booking, increasing transparency.

Lessons Learned

Uber's case demonstrates the double-edged nature of AI-driven dynamic pricing. It can balance markets efficiently, but, without transparency, dynamic pricing risks eroding customer trust. The lesson: dynamic pricing must be paired with clear communication and safeguards.

Expanded Case Example: E-Commerce and Retail

The Challenge

Amazon manages one of the world's largest online marketplaces, with millions of SKUs (stock-keeping units) and constant competition from other retailers. Manually setting or adjusting prices at this scale is impossible. Without dynamic systems, Amazon risked losing sales to competitors or failing to optimize margins in fast-moving categories.

The AI Solution

Amazon implemented AI-driven dynamic pricing algorithms that adjust prices on millions of products multiple times per day. The system analyzes:

- Competitor prices across marketplaces.
- Customer browsing and purchase histories.
- Inventory levels and restocking times.
- Seasonal and real-time demand shifts.

For example, if demand spikes for a household item, the system raises prices incrementally; if a competitor lowers their price, Amazon's algorithm often responds within hours.

The Results

- Studies estimate that Amazon reprices products every 10 minutes on average (Chen et al., 2016).
- Dynamic pricing is estimated to drive billions in additional revenue annually, helping Amazon maintain leadership in online retail.
- Competitors such as Walmart and Target have since adopted similar systems to remain competitive.

Lessons Learned

Amazon demonstrates the scalability of AI pricing. While humans cannot monitor millions of SKUs in real time, AI makes it possible to balance competitiveness and profitability at scale. However, Amazon's dominance has also raised regulatory questions about fairness and market power in algorithmic pricing.

The Balance Between Profitability and Trust

Dynamic pricing can significantly improve profitability, but it also carries risks:

- Customer backlash if prices are perceived as arbitrary or exploitative.
- Regulatory scrutiny in industries where price discrimination may raise fairness concerns.
- Brand erosion if customers feel loyal customers are penalized with higher prices.

The key is transparency and fairness. Communicating the reasons behind price changes (e.g., higher demand during peak times) and providing customer safeguards (such as caps or loyalty-based discounts) can help build acceptance.

📌 Guiding Questions for Reflection

- Where in my business could dynamic pricing improve revenue and efficiency (e.g., perishable inventory, peak demand periods)?
- What data sources (competitor pricing, real-time demand, contextual signals) could AI leverage for smarter pricing?
- How do I balance maximizing profit with maintaining customer trust?
- Do my current systems allow for transparent communication of dynamic pricing strategies?

Personalized Pricing and Offers

Personalization is one of the most powerful applications of AI in sales and pricing strategy. Instead of offering uniform discounts or flat pricing tiers, AI allows businesses to tailor offers to specific customers, micro-segments, or even individuals, based on their behavior, preferences, and willingness to pay.

A McKinsey (2021) study found that companies using advanced personalization techniques generate 40% more revenue from these efforts than their peers. At the same time, personalization must be managed carefully — if customers perceive pricing as unfair or discriminatory, trust can erode quickly.

How AI Powers Personalized Pricing

- Segmentation at scale: Machine learning creates thousands of micro-segments based on customer demographics, purchase history, and browsing behavior.
- Individualized offers: Dynamic coupons, loyalty perks, or discounts tailored to each customer.
- Predictive incentives: AI forecasts when a customer is likely to buy and offers incentives at the right time (e.g., before churn or during cart abandonment).
- Cross-sell and upsell optimization: Personalized bundles or add-on recommendations increase order value.

Expanded Case Example: Amazon's Personalized Promotions

The Challenge

As the world's largest online retailer, Amazon needed to increase conversion rates and basket size without resorting to blanket discounting. Traditional promotions wasted margin by offering discounts to customers who would have purchased anyway, while missing opportunities to convert those on the fence.

The AI Solution

Amazon employs AI-powered recommendation and personalization systems that:

- Analyze purchase history, search patterns, and browsing data.
- Deliver personalized coupons, promotions, and product bundles tailored to the individual.
- Use predictive analytics to time offers. For example, sending a discount on printer ink when past purchasing patterns suggest a refill is due.

The Results

- Personalized recommendations account for 35% of Amazon's total sales (Smith & Linden, 2017).
- Conversion rates increased significantly as offers aligned with customer intent.
- Amazon reduced unnecessary discounting, protecting margins while increasing revenue.

Lessons Learned

Amazon demonstrates that AI-powered personalization can increase both top-line growth and profitability. The key is using behavioral data not just to push products, but to anticipate real customer needs with relevant offers.

Expanded Case Example: Starbucks Rewards and DeepBrew

The Challenge

Starbucks needed to keep customers engaged in an increasingly crowded food and beverage market. Blanket promotions like "buy one, get one free" eroded margins and did little to drive loyalty beyond the initial purchase.

The AI Solution

Starbucks launched DeepBrew, an AI engine that powers its loyalty app and personalized marketing. DeepBrew analyzes customer purchase history, location, time of day, and even weather conditions to deliver individualized offers. For example:

- A customer who frequently buys lattes may receive a coupon for a new seasonal latte.
- On a hot afternoon, the app might suggest a cold brew.
- Rewards are personalized to maximize engagement, nudging customers toward higher-value items.

The Results

- Starbucks reported that AI-driven personalization helped increase loyalty program membership to over 30 million active users in the U.S. (Forbes, 2020).
- Personalized offers significantly boosted average ticket size and repeat visits.
- Starbucks deepened its competitive moat by embedding personalization into its customer experience.

Lessons Learned

Starbucks illustrates how AI can transform loyalty programs into powerful personalization engines. By tailoring offers to individual customers, businesses can build loyalty while simultaneously increasing sales and margins.

Expanded Case Example: Kroger's Loyalty Data Platform

The Challenge

Kroger, one of the largest U.S. supermarket chains, faced the challenge of customer loss and promotion inefficiency. Traditional weekly flyers and generic coupons did not resonate with increasingly digital-savvy shoppers who expected personalization similar to e-commerce experiences.

The AI Solution

Kroger leveraged its transaction data from the Kroger Plus loyalty program and, powered by AI machine learning, was able to:

- Analyze millions of loyalty card transactions and online behaviors.
- Deliver personalized coupons and discounts via app, email, and print.
- Predict customer needs and stock preferences to ensure relevant offers.

The Results

- Customers received offers with redemption rates several times higher than generic promotions.
- Kroger strengthened customer retention by building trust that offers were truly relevant.
- The program generated measurable ROI by reducing wasted discounts.

Lessons Learned

Kroger shows how personalization can transform even traditional industries like grocery retail. With the right AI systems, brick-and-mortar retailers can deliver digital-style personalization and win loyalty in competitive markets. Kroger went so far as to create 84.51°, a company that assists other retail organizations to understand shopper's behavior using this retail data science.

The Ethical Tightrope

While personalized pricing can unlock significant revenue gains, it raises ethical and regulatory concerns. Price discrimination based on personal characteristics, such as charging higher prices to wealthier customers, may backfire depending on the variables involved. The key is ensuring personalization is perceived as fair, relevant, and beneficial to the customer, rather than manipulative.

📌 **Guiding Questions for Reflection**

- How personalized are my current offers and promotions?
- Am I leaving revenue on the table with broad, generic discounts?
- How could AI predict the right moment to send an offer and to whom?
- What safeguards are in place to ensure personalization is fair and builds trust?

AI in Sales Forecasting

Forecasting has always been a cornerstone of business planning. Accurate predictions of customer demand help companies manage inventory, allocate resources, and plan promotions. Yet traditional forecasting methods, often based on spreadsheets, historical averages, or limited statistical models, struggle to capture the complexity and volatility of modern markets. Add in unpredictable consumer behaviors, global supply chain disruptions, and external shocks (such as pandemics or inflationary spikes) and forecasting with static models becomes increasingly unreliable. This is where AI creates transformational value, offering predictive power at a scale and granularity unavailable to human forecasters.

How AI Improves Sales Forecasting

AI-powered forecasting systems leverage machine learning, predictive analytics, and external data integration to create more accurate and adaptive projections:

- Granular demand prediction: AI can forecast demand at the SKU, store, or even customer level.
- Incorporating external signals: Weather patterns, social media sentiment, economic data, and competitor actions feed into models.
- Adaptive learning: Unlike static models, AI improves continuously as it ingests new data.
- Scenario modeling: AI can simulate "what-if" situations, such as price changes, competitor promotions, or supply chain disruptions.

McKinsey (2020) reports that companies adopting AI forecasting see forecasting errors reduced by 30–50%, enabling better planning and significant cost savings.

Expanded Case Example: Walmart's AI Forecasting System

The Challenge

As the world's largest retailer, Walmart manages over 100,000 products across thousands of locations. Traditional forecasting methods could not keep pace with the complexity of managing demand across different regions, seasons, and customer demographics. Even small forecasting errors resulted in overstocking, stockouts, and inefficiencies in supply chain planning.

The AI Solution

Walmart deployed AI-driven demand forecasting systems that:

- Analyzed real-time sales data from thousands of stores.
- Incorporated external signals such as weather forecasts, holidays, and regional events.
- Used machine learning models to predict demand not just at the national level, but at the store-by-store level.

The system also adjusted dynamically, updating forecasts daily as new data flowed in.

The Results

- Forecasting accuracy improved dramatically, reducing both overstocking and stockouts.
- The company achieved significant supply chain efficiency gains, lowering logistics costs while improving product availability.
- Customers benefited from fewer "out of stock" frustrations, improving satisfaction and loyalty.

Lessons Learned

Walmart's system demonstrates that AI forecasting isn't just about efficiency. AI forecasting directly improves customer experience and satisfaction by ensuring shelves are stocked with what customers want, when they want it.

Expanded Case Example: Unilever's AI Forecasting

The Challenge

Unilever, a global consumer goods giant with products sold in over 190 countries, faced challenges forecasting demand across diverse markets. With unpredictable customer preferences and volatile supply chains, the company needed more agile, data-driven forecasting methods.

The AI Solution

Unilever implemented AI-based demand forecasting tools, integrating internal sales data with external data sources such as weather, local holidays, and even social media sentiment. The system used advanced machine learning models to forecast demand at the product and market level.

The Results

- Forecasting errors dropped significantly, reducing wasted inventory.
- Supply chain agility improved, allowing Unilever to respond quickly to unexpected demand spikes.
- The company reported measurable reductions in working capital tied up in excess stock.

Lessons Learned

Unilever's experience highlights that AI forecasting creates agility, not just accuracy. By integrating external data streams, companies can adapt to fast-changing consumer demand and global uncertainties.

Expanded Case Example: Coca-Cola's Demand Forecasting

The Challenge

Coca-Cola, with its vast beverage portfolio, faced challenges predicting demand at the regional and seasonal level. For example, demand for sodas and bottled water varied widely depending on weather conditions, promotional campaigns, and cultural events.

The AI Solution

Coca-Cola deployed machine learning models that combined:

- Point-of-sale (POS) data from retailers.
- Local weather data (e.g., spikes in bottled water demand during heatwaves).
- Marketing campaign calendars and promotional activities.

The system allowed Coca-Cola to adjust production and distribution in real time, aligning supply with demand more effectively.

The Results

- Improved demand forecasting reduced both waste and missed sales opportunities.
- Coca-Cola increased on-shelf availability, especially in emerging markets where supply chain complexity was highest.
- Operational efficiency gains translated into cost savings across logistics and distribution.

Lessons Learned

Coca-Cola's system illustrates that AI forecasting does not just cut costs. AI forecasting creates competitive advantage by ensuring product availability where and when customers expect it.

📌 **Guiding Questions for Reflection**

- How accurate are my organization's current sales forecasts, and what costs result from errors?
- What external signals (weather, social media, economic trends) could AI incorporate to improve predictions?
- Could AI forecasting improve both operational efficiency and customer satisfaction?
- How agile is my forecasting process — can it adapt in real time as conditions change?

Revenue Management and Profit Optimization

Revenue management is the art and science of selling the right product, to the right customer, at the right time, for the right price. Traditionally, revenue management relied on spreadsheets, static forecasting, and broad discounting rules. These methods worked in relatively stable markets but struggle in today's world of volatile demand, complex supply chains, and empowered consumers. AI is transforming revenue management into a real-time optimization engine. By analyzing vast data streams, including historical sales, inventory levels, market demand, competitor actions, and customer behavior, AI systems continuously recommend pricing, promotions, and product mix strategies that maximize both short-term revenue and long-term profitability.

How AI Enhances Revenue Management

- Real-time adjustments: AI updates pricing and inventory allocation instantly in response to market shifts.
- Promotion optimization: Machine learning models evaluate which discounts generate incremental sales vs. margin erosion.
- Profit leakage detection: AI detects discount abuse, unprofitable contracts, or channel inefficiencies.
- Customer lifetime value (CLV) optimization: Beyond single transactions, AI prioritizes long-term profitability by tailoring offers to maximize retention and upsell opportunities.

The goal is not just to increase sales volume but to maximize profitability across the entire revenue cycle.

Expanded Case Example: Marriott International's Revenue Management System

The Challenge

Marriott operates thousands of hotels globally, each with different demand drivers, from leisure travel to business conferences. Traditional revenue management systems, while effective at a high level, struggled to optimize room pricing, promotions, and occupancy across such a complex portfolio.

The AI Solution

Marriott adopted an AI-powered revenue management system that:

- Forecasts demand by location, room type, and booking channel.
- Adjusts room prices dynamically based on real-time demand drivers such as holidays, events, and competitor rates.
- Recommends promotional strategies tailored to local markets.
- Integrates loyalty program data to identify opportunities to improve lifetime customer value.

The Results

- Marriott improved revenue per available room (RevPAR) by several percentage points, significantly impacting profitability.
- Forecasting accuracy improved, reducing reliance on heavy discounting to fill rooms.
- AI-powered insights enabled Marriott to maintain global pricing consistency while adapting locally.

Lessons Learned

Marriott's experience shows that AI revenue management is not only about pricing rooms, it is about strategically balancing occupancy, profitability, and customer loyalty at scale.

Expanded Case Example: Delta Airlines Yield Management

The Challenge

Airlines pioneered revenue management but faced increasing complexity in optimizing fares across thousands of routes, seat classes, and booking windows. Legacy systems were effective but often missed micro-trends in passenger demand and competitor behavior.

The AI Solution

Delta Airlines implemented AI-driven yield management systems capable of:

- Forecasting demand at a granular level by route, season, and booking behavior.
- Dynamically adjusting fares across multiple cabin classes.
- Incorporating real-time competitor price shifts and external signals such as weather or major events.
- Suggesting upsell opportunities, such as premium seating or bundled services.

The Results

- Increased overall load factor (percentage of seats filled) while maintaining profitability.
- Improved fare segmentation, allowing Delta to capture more willingness-to-pay across different customer groups.
- Enhanced operational agility, responding faster to changes in global travel demand.

Lessons Learned

AI enables airlines like Delta to go beyond static fare buckets, creating adaptive, data-driven yield systems that maximize both revenue and customer satisfaction.

Expanded Case Example: Netflix and Subscription Optimization

The Challenge

As a subscription-based business, Netflix's profitability depends not only on attracting new customers but also on retaining existing ones. Traditional pricing strategies for subscription services were blunt, offering limited differentiation and creating risks of customer loss.

The AI Solution

Netflix applied AI to optimize both pricing and product bundles. The system:

- Analyzes customer viewing habits, engagement patterns, and customer loss risks.
- Tests price sensitivity across different markets and demographics.
- Predicts which customers might respond positively to upgrades (e.g., premium HD plans) vs. which might leave if prices rise.

The Results

- AI-enabled testing allowed Netflix to roll out tiered pricing models (basic, standard, premium) optimized for different customer segments.
- Improved customer loss management by identifying at-risk customers and tailoring offers to retain them.
- Increased profitability while maintaining customer trust by balancing price changes with clear value communication.

Lessons Learned

Netflix's use of AI demonstrates that revenue management extends beyond physical inventory. In subscription businesses, AI helps balance customer acquisition, retention, and pricing strategies to maximize lifetime value.

The Strategic Value of AI Revenue Management

AI revenue management systems shift the focus from short-term pricing tactics to long-term profitability strategies. They enable businesses to capture untapped revenue, minimize waste, and improve customer loyalty through smarter offers. However, companies must also manage risks of algorithmic opacity — ensuring that revenue strategies remain transparent, explainable, and aligned with customer expectations.

- How is revenue currently managed in my organization — through manual oversight, static systems, or AI-driven platforms?
- Could AI identify profit leakages in discounts, contracts, or promotions?
- How do we measure success: short-term sales volume or long-term profitability?
- How transparent are our revenue optimization strategies to customers and stakeholders?

Sales Enablement with AI

Sales enablement is about equipping sales teams with the tools, insights, and processes needed to close deals effectively. Traditionally, this involved Customer Relationship Management (CRM) systems for tracking customer interactions, sales playbooks, and training programs. While helpful, these tools were often static, backward-looking, and heavily reliant on manual data entry, leaving sales representatives bogged down in administration rather than focusing on customer relationships. AI is transforming sales enablement by turning CRMs into predictive, proactive, and personalized platforms. Instead of just recording customer data, AI-enhanced systems analyze it, providing recommendations, prioritizing leads, and even coaching sales reps in real time. Many sales organizations are now augmenting traditional sales playbooks with AI-guided selling solutions, reflecting a profound shift in how sales teams operate.

How AI Powers Sales Enablement

- Predictive lead scoring: Machine learning models can assess which prospects are most likely to convert, based on past behavior, demographics, and engagement data.
- CRM intelligence: AI-enhanced CRMs (like Salesforce Einstein and HubSpot AI) suggest insights about customer intent, next-best actions, and deal risks.

- Real-time sales coaching: Conversation intelligence platforms analyze calls and meetings, suggesting improvements in tone, timing, and messaging.
- Personalized sales content: Generative AI tools help reps create tailored pitches, proposals, and follow-ups aligned with customer needs.

The goal is to shift sales reps from data entry clerks to strategic advisors empowered by the predictive insights AI can offer.

Expanded Case Example: Salesforce Einstein

The Challenge

CRM systems have long been criticized for being "glorified filing cabinets." Sales reps often spent hours inputting data with little immediate return, and managers lacked forward-looking insights. Salesforce, the world's largest CRM provider, recognized the need to make CRMs more intelligent and predictive.

The AI Solution

Salesforce introduced Einstein AI, an embedded AI engine within its CRM platform. Einstein analyzes customer interactions, emails, call notes, and past deals to provide:

- Predictive lead scoring: Highlighting which leads are most likely to convert.
- Opportunity insights: Identifying deals at risk of stalling.
- Next-best action recommendations: Suggesting when to reach out, what to say, and which channel to use.
- Automated data capture: Reducing the manual burden on sales reps.

The Results

- Companies reported improved conversion rates by focusing efforts on high-scoring leads.

- Sales reps spent more time selling and less time on administrative tasks.
- Managers gained clearer visibility into pipeline health, enabling data-driven coaching.

Lessons Learned

Salesforce Einstein illustrates how embedding AI into existing work-flows enhances adoption. The key lesson: AI must deliver actionable insights in real time to truly empower sales teams. Dashboards alone won't do the trick.

Potential Solution: Gong's Conversation Intelligence

The Challenge

Sales calls and meetings are rich sources of customer insight, but most organizations lacked the capacity to analyze them systematically. Valuable information about objections, competitor mentions, and buying signals often went unnoticed.

The AI Solution

Gong, a conversation intelligence platform, uses AI to record, transcribe, and analyze sales calls. The system provides:

- Real-time coaching: Suggesting adjustments in tone, pacing, and talk-to-listen ratios.
- Deal intelligence: Identifying red flags such as stalled decision-makers or negative sentiment.
- Competitive insights: Highlighting when competitors are mentioned during conversations.

The Results

- Companies using Gong reported shorter sales cycles and higher close rates.
- Sales managers gained unprecedented visibility into rep performance, enabling targeted coaching.

- AI insights revealed common customer objections, informing both product development and marketing.

Lessons Learned

Gong's case shows that AI not only improves individual rep performance but also creates organizational learning loops, where insights from one call inform the broader sales strategy.

Potential Solution: HubSpot AI

The Challenge

Small and mid-sized businesses (SMB) often lacked the resources to deploy complex AI tools for sales enablement. HubSpot sought to bring accessible AI-powered sales enablement to a broader audience.

The AI Solution

HubSpot integrated AI features into its CRM platform, offering:

- Lead scoring algorithms based on engagement, email opens, and website behavior.
- Content personalization tools to recommend the most effective templates for outreach.
- Forecasting models that predict revenue based on pipeline activity.
- Chatbots to qualify leads before passing them to human reps.

The Results

- SMBs using HubSpot's AI tools saw increased sales efficiency without needing dedicated data science teams.
- Predictive insights helped reps prioritize effectively, reducing wasted effort.
- Automated lead qualification improved both customer experience and sales outcomes.

Lessons Learned

HubSpot demonstrates that AI sales enablement isn't only for enterprise-level organizations. With user-friendly design, even smaller firms can harness AI for predictive selling and efficiency gains.

The Future of AI-Enabled Sales

AI is turning sales into a science as much as an art. The combination of predictive analytics, real-time coaching, and generative content creation means sales reps can be more effective, managers more strategic, and customers better served. However, the human element remains critical: empathy, creativity, and trust are still the foundation of successful selling. AI's role is to augment, not replace these qualities.

📌 **Guiding Questions for Reflection**

- How much time do my sales reps spend on data entry versus selling?
- Could predictive lead scoring help prioritize efforts and shorten sales cycles?
- Do managers have real visibility into sales conversations and pipeline risks?
- How can AI be integrated into existing workflows without overwhelming adoption?

Closing Thoughts

Pricing and sales optimization have always been central to business performance, but artificial intelligence is reshaping them into something far more powerful. What used to be a reactive discipline of adjusting spreadsheets and reacting to market shifts has become a proactive, predictive, and adaptive capability.

Through dynamic pricing, personalized offers, forecasting, and AI-enabled sales enablement, companies are discovering that revenue management is no longer just about maximizing margins, it's about creating fairer, more transparent, and more tailored experiences for customers. At the same time, AI's ability to process vast datasets at speed means businesses can react to fluctuations in demand, supply chain shocks, and competitive pressures with greater agility than ever before.

But as we explored, the ethical dimension cannot be ignored. The risks of over-personalization, bias in algorithms, and consumer backlash against opaque practices (as seen in Uber's surge pricing controversy or Amazon's marketplace scrutiny) underscore the need for businesses to strike a balance between profitability and fairness. Transparency, explainability, and safeguards against unintended consequences are no longer optional — they are strategic necessities.

For executives, the path forward involves more than just adopting AI tools. It requires a mindset shift: viewing pricing and sales not only as levers of financial performance but as moments of trust and relationship-building with customers. Companies that succeed will be those that harness AI's precision and scale while grounding their strategies in principles of fairness, accountability, and customer-centricity.

In the long run, AI-driven pricing and sales optimization will not simply separate the winners from the laggards. AI-driven pricing and sales optimization will eventually redefine how value itself is created and shared in the marketplace. Those who embrace this transformation thoughtfully will set the new standards for both profitability and trust in the AI-powered economy.

Please refer to the below appendices located at the end of the book for chapter-specific resources. However, be aware that inclusion in the appendices should not be considered an endorsement by the author for any individual commercial product.

APPENDIX O: AI-Powered Pricing & Sales Optimization Platforms

APPENDIX P: AI-Driven Pricing & Sales Personalization Platforms

APPENDIX Q: AI-Powered Sales Forecasting Platforms

APPENDIX R: AI-Powered Revenue Management & Profit Optimization Tools

CHAPTER 9
BALANCING PROFITABILITY WITH FAIRNESS

AS AI TRANSFORMS pricing and sales optimization, organizations face a delicate balancing act: how to maximize profitability without compromising fairness and trust. While AI-driven systems can uncover revenue opportunities, improve efficiency, and personalize offers, they also introduce risks of perceived or actual unfairness. Customers today are more data-savvy and quick to react to perceived exploitation. A PwC survey (2021) found that 88% of consumers said trust was a deciding factor in whether they buy from a brand, and 71% said they would stop doing business with a company that they no longer trusted. This makes fairness in pricing not only an ethical imperative but also a strategic business necessity.

The tension lies in AI's efficiency. Machine learning models are optimized for outcomes such as revenue maximization, margin growth, or conversion rates. Left unchecked, these systems may adopt strategies that unintentionally disadvantage certain groups, exploit peak demand, or create opaque pricing that customers do not understand. Consider the controversies around Uber's surge pricing during emergencies or the scrutiny of Amazon's algorithmic repricing in its marketplace. Both examples highlight the risk of short-term optimization clashing with long-term reputation. Even when technically efficient, algorithms that appear to exploit consumers can generate backlash, regulatory intervention, and lasting damage to customer relationships.

At the same time, regulators are stepping in. The European Union's upcoming AI Act and the U.S. Federal Trade Commission's guidance on "algorithmic fairness" place growing obligations on companies to ensure transparency and avoid discriminatory practices (FTC, 2021; European Commission, 2023). Businesses that fail to integrate fairness into their AI strategies risk not only reputational damage but also legal and financial penalties.

Ultimately, the ethical tightrope is about finding equilibrium. AI must be designed and governed in ways that balance profitability, compliance, and customer trust. Companies that succeed in this balancing act will not only avoid risks but also differentiate themselves in the marketplace as brands that combine intelligence with integrity.

📌 Guiding Questions for Reflection

- How important is customer trust in sustaining my organization's long-term profitability?
- Could any of our AI-driven pricing or sales strategies be perceived as exploitative?
- Are we proactively monitoring for fairness and bias in our algorithms?
- What role should transparency play in communicating AI-driven pricing to customers?

Risks of AI-Driven Pricing and Sales

AI brings enormous potential to transform pricing and sales optimization, but it also introduces ethical, reputational, and operational risks that organizations must address. If left unchecked, algorithmic systems can create outcomes that are technically efficient but socially unacceptable or legally problematic.

- **Perceived Unfairness**

One of the most significant risks is that customers may perceive AI-driven pricing as unfair or manipulative. For example, ride-hailing customers frequently expressed frustration with surge pricing, feeling they were being penalized during peak demand or emergencies. Even if the logic of surge pricing is to balance supply and demand, perception often overrides explanation, leaving lasting reputational damage.

Similarly, research shows that when customers discover they are being charged different prices for the same product, trust erodes quickly, even if the differences stem from legitimate market conditions (Martin & Murphy, 2017).

- **Discriminatory Outcomes**

AI algorithms are only as unbiased as the data on which they are trained. If training data reflects historical inequalities, algorithms can inadvertently perpetuate or even amplify discrimination. For instance:

- An AI pricing system might offer lower discounts to customers in lower-income neighborhoods, assuming lower price sensitivity.
- Loan pricing or credit offers could unintentionally correlate with race or gender if historical data encoded those biases.

These risks are not just reputational. These risks may raise serious legal and compliance issues, particularly under anti-discrimination and consumer protection laws.

- **Exploitation in High-Demand Scenarios**

Another risk is over-optimization during crises or emergencies. Dynamic pricing systems may raise prices significantly when demand spikes, leading to accusations of price gouging.

- During natural disasters, reports of bottled water being sold at inflated prices via online marketplaces led to public outrage and regulatory warnings (New York Times, 2017).
- In healthcare, concerns have been raised about AI pricing models used by insurers to set premiums that disadvantage vulnerable populations (Obermeyer et al., 2019).

While AI may optimize revenue in the short term, such practices can severely harm customer trust and invite regulatory intervention.

- **Opacity and Lack of Explainability**

Many AI pricing models operate as "black boxes," making decisions that even business leaders struggle to explain. This opacity undermines transparency and accountability. When customers do not understand why they are paying a certain price, they are more likely to assume manipulation. For example, studies of Amazon Marketplace repricing algorithms revealed opaque patterns of rapid price adjustments, sparking concerns about anti-competitive behavior and algorithmic collusion (Chen et al., 2016). Regulators increasingly view explainability as a requirement for compliance under frameworks like the EU's AI Act.

The Bottom Line

While AI can unlock new levels of profitability, these risks demonstrate that short-term efficiency gains can create long-term liabilities if fairness, transparency, and accountability are not built into the system. Companies that fail to recognize these risks may not only alienate customers but also attract legal and regulatory consequences.

📌 Guiding Questions for Reflection

- Could my AI pricing system be perceived as exploitative or unfair by customers?
- Are we monitoring for potential bias or discrimination in pricing and sales algorithms?
- How explainable and transparent are our AI-driven pricing models?
- Do we have safeguards in place to prevent "gouging" during crises or demand spikes?

Regulatory and Legal Considerations

As AI increasingly drives pricing and sales strategies, businesses face a growing set of legal and regulatory obligations. How much to charge, when, and to whom, was once considered a commercial decision. These decisions are now under scrutiny from consumer protection agencies, competition authorities, and data regulators worldwide. The central concern is that AI-driven pricing may result in unfair, discriminatory, or opaque outcomes, undermining consumer trust and potentially violating laws. To navigate this environment, organizations must understand the evolving global regulatory landscape and design systems that prioritize transparency, fairness, and compliance.

European Union: GDPR and the AI Act

The General Data Protection Regulation (GDPR), enacted in 2018, places restrictions on automated decision-making and profiling that "significantly affect" individuals (European Commission, 2018). Under Article 22, consumers have the right not to be subject to decisions made solely by automated systems, unless certain safeguards (such as human oversight) are in place. This means that AI-driven pricing systems that automatically assign different offers or premiums could fall under GDPR obligations. Companies must be prepared to explain how decisions are made and offer avenues for customer recourse.

Looking ahead, the EU Artificial Intelligence Act (AI Act), proposed in 2021, introduces a risk-based framework for regulating AI. Pricing and sales algorithms are not explicitly banned but may be classified as "high-risk systems" if they affect consumer rights. Obligations include:

- Transparency requirements: Customers must be informed when interacting with AI-driven systems.
- Bias monitoring: Companies must mitigate discriminatory outcomes.
- Explainability: AI pricing systems must be auditable and understandable to regulators.

Failure to comply could lead to fines of up to €30 million or 6% of global turnover, making compliance a strategic as well as legal necessity (European Commission, 2023).

United States: FTC and Antitrust Scrutiny

In the United States, the Federal Trade Commission (FTC) has taken a strong stance on algorithmic pricing and fairness. In 2021, the FTC issued guidance urging companies to ensure that AI systems are "truthful, fair, and equitable" and warned of enforcement against discriminatory or deceptive practices (FTC, 2021).

Key U.S. regulatory considerations include:

- Price discrimination laws: Under the Robinson-Patman Act, companies may face scrutiny if AI pricing disproportionately disadvantages certain customer groups.
- Antitrust enforcement: Concerns have been raised about algorithmic collusion, where pricing algorithms learn to coordinate without explicit agreements between competitors. The FTC and Department of Justice (DOJ) are monitoring cases where AI repricing systems may lead to anti-competitive outcomes.
- Consumer protection: Companies must avoid "dark patterns," meaning manipulative digital interfaces, which may be amplified by AI-driven personalization.

Other Jurisdictions

- United Kingdom: The Competition and Markets Authority (CMA) has launched inquiries into algorithmic pricing practices, emphasizing transparency and fairness (CMA, 2021).
- Asia-Pacific: Countries such as Singapore and Japan are issuing AI governance frameworks that stress human-centric AI and fairness in consumer markets.
- Global convergence: Organizations like the OECD are calling for international coordination on algorithmic accountability, recognizing the global nature of e-commerce and digital platforms.

Case Example: Amazon Marketplace

Amazon has faced scrutiny on multiple fronts regarding its algorithmic pricing practices. Studies found that Amazon's repricing algorithms adjusted prices so frequently that they occasionally created conditions resembling tacit collusion, raising antitrust concerns (Chen et al., 2016). Regulators in both the U.S. and EU have investigated whether such systems may undermine competition. While Amazon has defended its systems as customer-focused, the case illustrates the growing expectation that companies must monitor and explain AI behavior proactively to regulators and the public.

The Compliance Imperative

The message from regulators is clear: AI pricing cannot be a black box. Organizations must:

- Document their AI models and decision-making processes.
- Build human oversight into pricing systems.
- Provide transparency to consumers on how offers are determined.
- Regularly audit for fairness, bias, and compliance with competition law.

Those who fail to act face not only legal risks but also reputational fallout in a marketplace where customers demand both fair value and fair treatment.

🎯 Guiding Questions for Reflection

- Do our AI-driven pricing systems fall under GDPR's rules on automated decision-making?
- Are we prepared for the EU AI Act's transparency and auditability requirements?
- Could our algorithms unintentionally create discriminatory outcomes or appear collusive?
- Do we have processes in place to explain pricing decisions to customers and regulators?

Principles for Fair AI Pricing and Sales

To balance profitability with fairness, organizations need clear principles and practices that ensure AI systems deliver value to customers as well as the business. The most successful companies are likely to be those that treat fairness not as a compliance obligation, but as a strategic asset that builds long-term trust, loyalty, and differentiation in competitive markets.

Below are four guiding principles that can help organizations design and implement AI pricing and sales systems responsibly.

1. Transparency

Customers are more likely to accept dynamic or personalized pricing when they understand how it works. Transparency involves:

- Clear communication: Explaining why prices may fluctuate (e.g., demand-based adjustments or loyalty-based discounts).
- Upfront disclosure: Providing customers with information before a transaction, as Uber does with estimated fare ranges to reduce surprise and backlash.

- Internal transparency: Documenting how AI systems make decisions to satisfy regulators and build organizational accountability.

Transparency doesn't mean exposing proprietary algorithms but ensuring customers feel pricing is predictable, fair, and understandable.

2. Equity

Equity means avoiding outcomes that unfairly disadvantage specific groups. Best practices should include:

- Bias auditing: Regularly testing models for disparate impacts across demographics, geography, or income levels.
- Inclusive data: Ensuring training datasets represent diverse populations, rather than perpetuating historical inequities.
- Equitable value exchange: Offering benefits (discounts, promotions, loyalty rewards) in ways that serve all customers, not only high-value segments.

Organizations that proactively address equity not only reduce legal risks but also strengthen brand reputation as inclusive and customer-focused.

3. Customer-Centric Design

AI pricing and sales strategies should focus on creating mutual value: customers should feel they benefit from personalization rather than being exploited. This involves:

- Designing offers that are timely, relevant, and helpful, such as reminders to repurchase consumables or personalized loyalty rewards.
- Ensuring personalization feels empowering rather than invasive — avoiding the "creepiness factor" where customers feel liked they are being stalked online.

- Embedding customer feedback loops into AI systems to continuously refine and align pricing strategies with consumer expectations.

Customer-centric design reframes AI from a tool of extraction to a driver of relationship building.

4. Safeguards and Human Oversight

AI pricing should never operate unchecked. Safeguards are necessary to prevent harm, particularly in sensitive contexts:

- Caps on dynamic pricing to avoid "gouging" during crises or emergencies.
- Human-in-the-loop oversight, ensuring managers can review and override automated decisions when necessary.
- Regular audits for fairness, explainability, and regulatory compliance.

These measures help organizations demonstrate responsible governance, reassuring both customers and regulators.

The Strategic Payoff

Companies that adopt these principles are not only less exposed to regulatory risk, but also more likely to build customer trust as a source of competitive advantage. As PwC (2021) notes, trust is increasingly a differentiator in consumer markets. Therefore, fairness in AI-driven pricing and sales can be a powerful trust-building mechanism.

🖈 Guiding Questions for Reflection

- Do we explain to customers how dynamic or personalized pricing works, in ways they can understand?
- Are we actively monitoring for bias and inequity in our AI-driven sales models?
- Does our personalization strategy create genuine value for customers, or could it be perceived as manipulative?

- What safeguards and human oversight mechanisms do we have in place to prevent harmful pricing outcomes?

Expanded Case Example: Uber Surge Pricing Backlash

The Challenge

As Uber scaled globally, one of its most pressing challenges was balancing real-time rider demand with driver supply. Traditional flat-rate pricing models left customers stranded during peak times, such as concerts, sports events, or bad weather, while drivers had little incentive to increase availability. Uber needed a way to dynamically balance the marketplace to ensure reliability.

However, the solution also carried significant risks: riders could interpret fluctuating prices as exploitative or unfair, especially in emergencies. The company would need to walk a fine line between efficiency and customer trust.

The AI Solution

Uber implemented an AI-driven surge pricing system. The algorithm monitored:

- Rider demand and trip request density in real time.
- Active driver supply and geographic availability.
- External contextual data such as traffic patterns and event schedules.

When demand exceeded supply, the algorithm raised fares dynamically. The higher prices served two purposes:

1. Incentivize drivers to move toward high-demand areas.
2. Ration demand by discouraging lower-priority trips, ensuring that riders most willing to pay could secure a car.

Uber also introduced upfront pricing to display estimated fares and surge multipliers before booking, an attempt at transparency after initial backlash (Cohen et al., 2016).

The Results

- Operational Efficiency: Surge pricing significantly improved the reliability of service during peak times by bringing more drivers online.
- Driver Earnings: Drivers earned more in high-demand periods, improving satisfaction and retention.
- Customer Backlash: Riders often perceived surge pricing as exploitative, particularly during crises. For example, during Hurricane Sandy in 2012, media outlets reported fares several times higher than normal, sparking accusations of "price gouging" (New York Times, 2017).
- Policy Adjustments: Facing reputational risk, Uber introduced caps on surge pricing during natural disasters and emergencies, aligning its practices with fairness expectations and regulatory guidance.

Lessons Learned

Uber's surge pricing illustrates both the power and peril of AI-driven pricing. On the one hand, it improved marketplace efficiency and driver supply. On the other, it highlighted that profit optimization must be balanced with public perception and ethics.

Key takeaways include:

1. Transparency is critical: Clear communication about why prices are rising helps reduce backlash.
2. Safeguards build trust: Caps on pricing during emergencies showed responsiveness to fairness concerns.
3. Reputation matters as much as efficiency: Even if an algorithm achieves operational goals, customer trust can erode if fairness is not prioritized.

Uber's experience underscores that dynamic pricing is not just a technical challenge but also a reputational and ethical one.

Expanded Case Example: Amazon Marketplace Scrutiny

The Challenge

Amazon operates the world's largest online marketplace, with millions of third-party sellers and constantly shifting competition. Managing prices manually at this scale is impossible. To remain competitive and maximize profitability, Amazon and its sellers increasingly relied on algorithmic repricing tools that automatically adjusted prices in response to competitors.

While these tools improved efficiency, they also created regulatory and ethical risks. Studies suggested that Amazon's algorithms might unintentionally facilitate "tacit collusion" — where independent sellers' pricing bots repeatedly undercut or shadow each other until prices aligned at higher levels, reducing competition (Chen et al., 2016). Regulators in both the U.S. and Europe began investigating whether algorithmic repricing practices harmed consumers or violated antitrust law.

The AI Solution

Amazon developed and deployed AI-driven dynamic repricing systems, both internally and for third-party sellers. These tools:

- Adjusted millions of product prices multiple times per day.
- Monitored competitor listings, customer demand signals, and inventory levels.
- Sought to optimize for the "Buy Box" — Amazon's default featured offer, which drives a majority of sales.

To address growing scrutiny, Amazon implemented compliance safeguards:

- Enhanced monitoring of third-party pricing behaviors.
- Policies against explicit price-fixing or collusion, with enforcement mechanisms.
- Public statements affirming that pricing algorithms were designed to benefit customers through competition.

The Results

- Operational Efficiency: Amazon maintained its reputation for offering competitive prices by enabling near-real-time repricing.
- Revenue Gains: Algorithmic repricing contributed significantly to sales growth and margin optimization, helping Amazon sustain leadership in e-commerce.
- Regulatory Scrutiny: Academic research and antitrust authorities raised concerns about potential consumer harm, including reduced competition and inflated prices in some product categories (Ezrachi & Stucke, 2017).
- Ongoing Investigations: While Amazon has not been found guilty of explicit collusion, regulators continue to monitor its algorithms and policies closely, setting precedents for global AI oversight in pricing.

Lessons Learned

Amazon's case demonstrates both the power and risk of algorithmic pricing at scale:

1. Efficiency must be balanced with compliance: Repricing tools created competitive advantages but also regulatory exposure.
2. Unintended consequences matter: Even without explicit coordination, algorithms can produce collusive outcomes if not carefully designed and monitored.
3. Proactive governance is essential: Monitoring, auditing, and transparent communication help mitigate both legal and reputational risks.

The Amazon case underscores a broader lesson: AI pricing systems can have market-wide impacts that go beyond individual companies, making governance and oversight critical to ensuring fair competition.

Building Trust Through Ethical AI

The previous sections have highlighted both the opportunities and risks of AI-driven pricing and sales. From Uber's surge pricing backlash to Amazon's algorithmic scrutiny, the lesson is clear: profitability without fairness undermines trust, and without trust, profitability is unsustainable.

To turn fairness into a strategic advantage, businesses must go beyond compliance and embed ethical AI principles into their operations. Trust is not simply a byproduct of efficiency; it must be designed, communicated, and continuously reinforced (PwC, 2021).

Actionable Strategies for Ethical AI

1. Embrace Explainable AI (XAI)

Customers and regulators increasingly demand transparency in algorithmic decisions. Explainable AI (XAI) ensures that pricing and sales recommendations are not "black boxes" but can be audited, explained, and justified.

- For customers: Provide clear messaging on why a price or offer was generated.
- For regulators: Maintain documentation of model design, training data, and testing processes.
- For managers: Enable dashboards that show "reason codes" for pricing decisions.

2. Build Fairness into Model Design

Fairness cannot be retrofitted; it must be embedded during development. Best practices include:

- Conducting bias audits before deployment.
- Using diverse datasets to avoid reproducing historical inequalities.
- Creating fairness metrics (e.g., equalized error rates across demographics) and monitoring them regularly.

3. Implement Guardrails and Human Oversight

AI should not operate without boundaries. Safeguards ensure that optimization does not slip into exploitation:

- Set caps on dynamic pricing to prevent "gouging" during crises.
- Use human-in-the-loop review for high-stakes or sensitive pricing decisions.
- Regularly audit algorithmic performance to detect drift or unintended effects.

4. Communicate Value, Not Just Price

Trust grows when customers feel that personalization benefits them, not just the business. Companies should:

- Frame AI-driven offers as helpful and relevant (e.g., loyalty perks, timely reminders).
- Provide customers with control options, such as opting out of personalized pricing.
- Use messaging that emphasizes mutual value: "We're offering you this discount because you're a valued customer."

5. Treat Fairness as a Brand Differentiator

Fair AI practices are more than risk mitigation — they can be leveraged as a competitive advantage. In crowded markets, brands that proactively highlight fairness, transparency, and ethical safeguards may stand out with customers and regulators alike.

Synthesis of Case Study Lessons

- Uber taught us that efficiency without transparency creates backlash; safeguards and clear communication are essential.
- Amazon demonstrated the dangers of algorithmic opacity and the need for proactive compliance and governance.
- Marriott, Delta, Netflix, and Starbucks showed that

personalization and optimization can succeed when designed with customer-centric principles.

The synthesis is clear: AI must serve both profitability and fairness. One without the other is unsustainable.

The Long-Term Payoff

Companies that integrate ethical AI practices will not only reduce regulatory and reputational risks but also build enduring trust with customers. In an era where consumers are increasingly skeptical of algorithmic decision-making, being known as a company that uses AI responsibly can become a key differentiator.

Trust, once earned, compounds over time. As Rawson et al. (2013) argue, customer experience is the ultimate competitive advantage — and in the AI era, fairness is at the heart of that experience.

🎯 **Guiding Questions for Reflection**

- Do we have explainability and auditability built into our AI systems?
- Are fairness and bias metrics monitored as rigorously as revenue metrics?
- How do we communicate the value of AI-driven offers to customers?
- Could we position our commitment to ethical AI as a brand strength?

Reflection and Discussion

Balancing profitability with fairness is one of the defining challenges of AI-driven pricing and sales. While algorithms can unlock powerful efficiencies and revenue gains, the risk of eroding trust is ever-present if fairness, transparency, and ethics are not prioritized. This closing section invites leaders to reflect on how the lessons of this chapter apply to their own organizations — not only as a matter of compliance, but as a strategic opportunity to strengthen brand trust.

📌 **Guiding Questions for Organizational Reflection**

1. **Trust and Customer Perception**
   - How important is customer trust in sustaining our organization's long-term profitability?
   - Could any of our AI-driven pricing or sales strategies be perceived as exploitative or unfair?
   - Have we tested customer reactions to personalized or dynamic pricing models?

2. **Bias and Fairness in AI Systems**
   - Are we auditing our AI systems for unintended bias or discriminatory outcomes?
   - Do our training datasets represent the diversity of our customer base?
   - How do we measure fairness — and are those metrics weighted as heavily as revenue outcomes?

3. **Transparency and Explainability**
   - Can we explain to customers, in simple terms, how pricing and sales recommendations are generated?
   - Do our internal teams (sales, legal, compliance) fully understand and trust our AI-driven systems?
   - How prepared are we to provide documentation and justification to regulators if required?

4. **Safeguards and Governance**
   - Do we have clear caps, rules, or override mechanisms to prevent harmful pricing outcomes (e.g., "gouging" during crises)?
   - Who is responsible for human oversight of AI pricing systems in our organization?
   - Are we conducting regular audits for compliance with GDPR, the EU AI Act, FTC guidance, or other local regulations?

5. **Strategic Opportunities**
   - Could we position fairness and transparency as differentiators in our industry?
   - How can our commitment to ethical AI become part of our brand identity?

○ What steps can we take today to build a culture of **responsible AI use** that aligns profitability with long-term trust?

Discussion Prompts for Teams

- *Scenario 1*: Imagine our AI system inadvertently raised prices during a regional emergency. How would we respond — to customers, regulators, and the media?
- *Scenario 2*: A regulator requests full documentation of how our AI models determine pricing. Could we provide it quickly and confidently?
- *Scenario 3*: A competitor begins publicly advertising their "fair AI" pricing practices. How might we respond to remain competitive?

Closing Thought

The organizations that will thrive in the AI era are those that recognize fairness as more than a constraint. Fairness will be recognized as a strategic advantage. Profitability may come from optimization, but enduring growth comes from trust. Leaders who embrace this dual imperative will not only future-proof their businesses against regulation and disruption but also build brands that customers respect, value, and remain loyal to over the long term.

CHAPTER 10
AI IN SUPPLY CHAIN AND RESOURCE ALLOCATION

IF FINANCE IS the lifeblood of an organization, then the supply chain is its backbone. It's what keeps products moving, services flowing, and customers satisfied. A strong backbone ensures agility and resilience. A weak supply chain can paralyze even the most innovative business. Yet, managing supply chains has never been more complex. Globalization, rising customer expectations, and frequent disruptions, from trade disputes to pandemics, have made it harder for businesses to balance efficiency with resilience. In a recent Deloitte survey (2022), 79% of executives reported that supply chain disruptions had a significant negative impact on their companies. Traditional approaches, built on historical averages and manual forecasting, simply cannot keep up with this volatility.

This is where AI comes in. AI allows companies to move from reactive management of problems after they occur, to enjoying predictive and adaptive supply chains that anticipate challenges before they happen. By analyzing massive datasets, AI can forecast demand more accurately, optimize inventory levels, reroute logistics in real time, and even detect risks hidden deep within supplier networks. Consider the difference in these approaches. Like with traditional sales forecasting in Chapter 6, traditional supply chains often resemble a rearview mirror, showing leaders what happened last month or last quarter. AI-enabled supply chains, on the other hand, function like a navigation system. They monitor real-time conditions, forecasting what's ahead, and recommending the best course of action. That's not just operational improvement. That is strategic advantage.

The importance of this shift became crystal clear during the COVID-19 pandemic. Companies with AI-enabled forecasting and adaptive logistics bounced back faster, while those dependent on static, manual systems struggled to secure critical resources (World Economic Forum, 2021). As McKinsey put it, AI-enabled supply chains are now a "must-have" capability for resilience, not a nice-to-have efficiency upgrade (McKinsey & Company, 2020).

In short, supply chains are no longer just about moving goods. They are about anticipation, adaptation, and resilience. And AI is the technology making that possible.

🖈 Reflection Questions

- How dependent are we on outdated, reactive supply chain practices?
- Do we view our supply chain as a strategic asset — or just a cost center?
- Could AI help us turn volatility into an opportunity rather than a threat?

Demand Forecasting and Inventory Optimization

If supply chains are the backbone of business, then demand forecasting is the nervous system. Demand forecasting signals when to stock up, when to pull back, and where to direct resources. Get it right, and products flow seamlessly to customers. Get it wrong, and the consequences are painful: stockouts that frustrate customers, overstocks that tie up cash, and waste that eats into profits. For decades, businesses relied on traditional forecasting methods like historical averages and seasonal adjustments. But in today's volatile environment, those models fall way short. Consumer behavior shifts rapidly, supply shocks ripple globally, and external factors like weather, social trends, or even a viral TikTok video can send demand soaring or crashing overnight.

This is where AI-driven forecasting shines. By analyzing massive datasets that might come directly from point-of-sale (POS) transactions or loyalty data, to external signals like weather forecasts, promotions, and social sentiment, AI can predict demand at a much more granular level. Instead of forecasting at the national or regional level, AI can generate SKU-level forecasts for individual stores or warehouses, adjusting predictions in near real time.

The payoff is significant. A McKinsey study found that AI-enabled demand forecasting can reduce forecasting errors by 30–50%, leading to a 2–6% increase in revenue and a 10–20% reduction in inventory costs (McKinsey, 2018). That's a huge win for both profitability and resilience.

Expanded Case Example: Walmart's AI-Driven Inventory System

The Challenge

Walmart operates over 10,000 stores worldwide, each carrying thousands of products. Traditional forecasting models struggled to account for regional nuances, such as weather patterns or local buying habits. Stockouts in one store and overstocks in another were common, leading to lost sales and wasted inventory.

The AI Solution

Walmart turned to AI-driven demand forecasting powered by machine learning models. These systems integrate POS data, online sales, weather information, regional events, and even local demographics. The AI continuously learns and updates predictions, helping Walmart anticipate shifts in demand down to the individual store and product level.

For example, if a cold front moves into the Midwest, the system predicts higher demand for heaters and winter apparel, and automatically adjusts inventory and replenishment orders. If social media buzz spikes for a new product, Walmart can adjust stock levels in real time to meet demand.

The Results

- Reduced stockouts across key product categories, improving customer satisfaction.
- Lower inventory carrying costs, as products were better matched to demand.
- Improved shelf availability boosted sales, particularly in grocery and seasonal categories.
- Enhanced agility helped Walmart respond more effectively to COVID-era disruptions.

Lessons Learned

Walmart's case shows that demand forecasting isn't just about efficiency. It's also about responsiveness. AI gave Walmart the ability to anticipate demand shifts faster than competitors, turning volatility into an advantage rather than a liability.

While industry giants like Walmart might appear to be financially more capable of creating their own solutions in-house, there are some commercially available platforms like Prediko and Monocle that appear to be making demand forecasting accessible and affordable for small to mid-sized e-commerce businesses without analytics teams.

📌 **Guiding Questions for Reflection**

- Do we still rely on historical averages for forecasting?
- How costly are stockouts or overstocks to our business today?
- Could AI give us the ability to forecast demand at a more granular level — down to the store, product, or even customer segment?

Logistics and Transportation Optimization

Getting products into customers' hands isn't just about having them in stock. It's also about moving them efficiently. Logistics and transportation are often the most expensive parts of the supply chain, sometimes accounting for up to 50% of total supply chain costs. Every mile driven, every truck half-full, and every inefficient delivery route adds up quickly. Traditionally, companies relied on static routing models or manual scheduling. Drivers followed predetermined routes that didn't always account for traffic conditions, fuel efficiency, or last-minute changes in demand. The result: wasted miles, higher emissions, and higher costs.

AI is transforming logistics by making it smarter and more adaptive. Machine learning algorithms analyze GPS data, weather conditions, fuel consumption, customer delivery windows, and even real-time traffic patterns. The goal isn't just efficiency, it's dynamic optimization. Routes can be recalculated on the fly, deliveries reprioritized mid-route, and resources reallocated as conditions change.

AI-driven logistics also has sustainability benefits. By cutting unnecessary miles and improving load efficiency, companies not only reduce costs but also lower carbon emissions. This is a win for both the bottom line and the environment.

Expanded Case Example: UPS's ORION Route Optimization

The Challenge

UPS delivers over 20 million packages per day. With such a massive volume, even small inefficiencies in routing translated into enormous costs. Traditional static routes often failed to adapt to daily variables like weather, traffic jams, or delivery cancellations. UPS needed a way to optimize routes dynamically, at scale.

The AI Solution

UPS developed ORION (On-Road Integrated Optimization and Navigation), an AI-powered system that analyzes vast amounts of data — including package destinations, traffic conditions, fuel consumption, and historical delivery times. ORION uses advanced algorithms to determine the most efficient sequence of stops for each driver, often recalculating routes in real time. A famous UPS innovation was the algorithm's preference for minimizing left turns (in right-driving countries), since left turns generally waste more fuel and time waiting at intersections.

The Results

- ORION saves UPS an estimated 100 million miles annually.
- This translates to 10 million gallons of fuel saved and reductions of 100,000 metric tons of CO_2 emissions per year (UPS, 2020).
- Cost savings are measured in the hundreds of millions of dollars annually.
- Driver productivity and on-time delivery rates improved significantly.

Lessons Learned

UPS's ORION shows the power of combining operational data with AI optimization. Logistics isn't just about getting packages from point A to point B — it's about doing it smarter, cheaper, and greener. AI-driven routing demonstrates how even small operational improvements can deliver massive scale benefits when applied across global networks.

📌 Guiding Questions for Reflection

- How much of our supply chain costs come from transportation and logistics inefficiencies?
- Could AI help us optimize routing, reduce emissions, and save money simultaneously?
- Are we still treating logistics as a cost center, or as a strategic advantage?

Supplier Risk Management

A supply chain is only as strong as its weakest link, which is often the supplier. From raw materials to finished goods, suppliers represent a huge share of risk. Natural disasters, geopolitical conflicts, financial instability, or even a single quality failure can ripple across an entire network. The 2011 Fukushima earthquake, for instance, disrupted global automotive and electronics supply chains for months, exposing how fragile supplier networks can be. Traditionally, supplier risk management relied on static assessments: financial audits, compliance certifications, or annual reviews. But these snapshots often failed to capture emerging risks. By the time a problem became visible, such as a bankrupt supplier, a delayed shipment, or a compliance violation, the damage was already done.

AI is changing the equation. Modern risk analytics platforms use machine learning and natural language processing (NLP) to continuously monitor signals from financial filings, news reports, weather data, social media, and even satellite imagery. These systems flag early-warning signs like deteriorating financial health, political instability in supplier regions, or shifts in consumer sentiment. Some tools also model the cascading impact of a disruption, estimating how one supplier failure might affect downstream production. This proactive approach allows companies to diversify sourcing, negotiate contingencies, or adjust production *before* the risk becomes a crisis. In today's volatile environment, supplier resilience is no longer optional; it's a strategic advantage.

Expanded Case Example: Siemens and AI-Driven Supply Chain Risk Analytics

The Challenge

Siemens operates one of the largest and most complex global supply chains, spanning tens of thousands of suppliers across multiple industries. Traditional supplier risk assessments were slow, manual, and reactive. Siemens faced growing challenges: global trade disruptions, cybersecurity threats, and sustainability requirements. The company needed a way to continuously monitor and assess risk across its vast supplier network.

The AI Solution

Siemens adopted AI-powered risk analytics platforms to track and evaluate suppliers in real time. These systems used:

- NLP algorithms to scan news, legal filings, and regulatory announcements for potential supplier issues.

- Predictive models to estimate the probability of delays, financial instability, or compliance violations.
- Risk scoring systems that ranked suppliers by vulnerability across dimensions like environmental, financial, and geopolitical risk.

AI also enabled Siemens to simulate disruption scenarios: *What happens if a supplier in Eastern Europe goes offline? How would this impact production in Asia or customer deliveries in the U.S.?* These digital "what-if" analyses informed contingency planning and inventory buffers.

The Results

- Siemens reduced supply chain blind spots by detecting risks months earlier than traditional systems.
- Procurement teams gained visibility into high-risk suppliers, enabling proactive mitigation.
- Supplier resilience became a measurable metric, supporting Siemens's broader sustainability and compliance goals.

Lessons Learned

Siemens's experience shows that supplier risk management is no longer about reacting to crises but instead, it's about anticipating them. AI empowers companies to monitor vast supplier ecosystems continuously, prioritize risks, and act before disruptions spiral. In doing so, risk management shifts from a defensive exercise to a strategic enabler of resilience.

📌 **Guiding Questions for Reflection**

- How much visibility do we really have into the risks within our supplier base?
- Are we monitoring suppliers continuously, or only through periodic reviews?
- Could AI-driven analytics help us move from reacting to disruptions to predicting them?

Resource Allocation and Workforce Scheduling

Getting the right products to the right place is only half the battle. To be successful, businesses also need the right people in the right roles at the right time. Workforce allocation has long been a tricky balancing act. Schedule too few employees, and customer service suffers. Schedule too many, and labor costs balloon. For industries like retail, hospitality, and healthcare, where margins are tight and demand fluctuates daily, scheduling inefficiencies can be devastating. Traditionally, managers created schedules manually, using spreadsheets or static rules. These methods often ignored critical factors like local demand surges, employee preferences, or compliance with labor laws. The result? Overtime costs, burnout, absenteeism, and frustrated workers who felt their personal needs weren't considered.

AI is transforming this problem into an opportunity. Modern workforce management platforms use machine learning to analyze customer demand patterns, historical sales, employee availability, and even local events. They can automatically generate optimal schedules that balance business needs with employee fairness. For managers, this means fewer headaches and better staffing alignment. For employees, it can mean more predictable schedules, reduced burnout, and higher engagement. In other words, AI is turning scheduling from a compliance exercise into a strategic lever for both efficiency and retention.

Expanded Case Example: Kronos (UKG) AI Scheduling in Retail

The Challenge

Large retailers face constant demand fluctuations, from weekend surges to seasonal shopping peaks. Store managers often spent hours creating schedules, only to end up either overstaffed during slow periods or understaffed during busy hours. Employees were frustrated by unpredictable shifts and frequent last-minute changes. High turnover and absenteeism were costly, and customer satisfaction suffered.

The AI Solution

Kronos (now Ultimate Kronos Group, UKG) implemented AI-powered workforce scheduling tools. These platforms integrated:

- Point-of-sale (POS) data to forecast customer traffic at the store and department level.
- Employee preferences and availability, reducing scheduling conflicts.
- Labor law and union compliance rules, ensuring legal adherence automatically.

The system generated optimized schedules that matched staffing levels to predicted demand while also taking fairness into account. Managers could adjust schedules quickly when demand shifted, with AI continuously learning from new data.

The Results

- Retailers using Kronos reported reduced labor costs while improving customer service.
- Employee satisfaction rose due to more predictable, fairer schedules.
- Turnover rates decreased, as workers felt their needs were respected.
- Managers saved hours of administrative time each week, focusing instead on customer-facing tasks.

Lessons Learned

Kronos's case highlights that AI-driven scheduling isn't just about cost-cutting — it's about building a more resilient, engaged workforce. By balancing efficiency with fairness, companies can improve both employee morale and customer experience, turning scheduling into a competitive advantage.

- Are we still creating schedules manually, ignoring real-time demand signals?
- How much are scheduling inefficiencies costing us in overtime, turnover, and poor service?
- Could AI-driven scheduling improve both employee satisfaction and operational performance?

Sustainability and Circular Supply Chains

Supply chains aren't just about cost and efficiency anymore. Supply chains are increasingly about sustainability. Customers, investors, and regulators are demanding that companies reduce their environmental impact, cut waste, and embrace circular economy principles. According to a PwC survey, 79% of consumers say they are more likely to buy from companies with a strong sustainability track record (PwC, 2021). Traditional supply chains have often been linear: extract, produce, use, and dispose. But this model produces enormous waste and leaves companies exposed to resource shortages and regulatory risks. The new frontier is the circular supply chain — one designed to reuse, recycle, and extend the life of resources.

Here, AI plays a pivotal role. Machine learning can:

- Optimize energy usage in factories and logistics.
- Predict when machines or vehicles will need maintenance, reducing downtime and waste.
- Monitor carbon emissions across the supply chain.
- Identify opportunities for reuse and recycling by analyzing product lifecycle data.

AI-enabled "digital twins" — virtual models of factories, supply chains, or products — allow businesses to simulate sustainability scenarios: *What if we reduce packaging by 20%? What if we shift sourcing to renewable suppliers?* This helps leaders balance financial, operational, and environmental goals simultaneously (Tseng et al., 2019).

Expanded Case Example: Unilever's AI-Enabled Sustainability Initiatives

The Challenge

As one of the world's largest consumer goods companies, Unilever faced immense pressure to cut its carbon footprint and embrace sustainable practices across its vast supply chain. The challenge was scale: Unilever sources from thousands of suppliers, manufactures in over 190 countries, and serves billions of consumers. Traditional sustainability reporting and manual monitoring were too slow and fragmented to drive meaningful change.

The AI Solution

Unilever turned to AI-powered supply chain management to embed sustainability into its operations. Key initiatives included:

- Predictive analytics to optimize energy use in manufacturing plants, lowering emissions.
- AI-driven logistics to reduce empty miles and fuel consumption in distribution.
- Satellite imaging and machine learning to monitor agricultural suppliers for deforestation risks.
- Digital twin technology to simulate resource use, packaging design, and waste reduction scenarios before rolling out changes at scale.

The Results

- Significant reductions in greenhouse gas emissions across manufacturing and logistics.
- Progress toward Unilever's goal of making all plastic packaging reusable, recyclable, or compostable.
- Improved supplier compliance with sustainability standards, thanks to AI-powered monitoring.
- Stronger brand reputation as a leader in sustainable business practices.

Lessons Learned

Unilever's case illustrates that sustainability isn't a trade-off against efficiency — it can be an enabler of long-term resilience and profitability. By using AI to monitor, optimize, and simulate, companies can make sustainability measurable and actionable. The lesson: when AI aligns supply chains with environmental goals, it creates value for customers, regulators, and shareholders alike.

🎯 **Guiding Questions for Reflection**

- Are sustainability goals fully integrated into our supply chain strategy, or treated as add-ons?
- Could AI help us track emissions, waste, and resource use in real time?
- How can digital twins or predictive models support our transition toward a circular supply chain?

Resilience in Times of Disruption

If the last few years have taught businesses anything, it's this: supply chain disruptions are no longer rare events — they are the new normal. From COVID-19 to the war in Ukraine to climate-related disasters, global supply networks have been under near-constant stress. According to McKinsey, companies can now expect supply chain disruptions lasting a month or longer to occur every 3.7 years on average (McKinsey & Company, 2020). Traditional supply chains were built for efficiency, not resilience. Lean inventory, single sourcing, and rigid logistics worked well in stable times but left companies vulnerable when shocks hit. The pandemic, for example, exposed how brittle just-in-time systems could be when factories shut down or borders closed.

AI offers a powerful way to shift from brittle efficiency to adaptive resilience. By integrating real-time data, such as shipping delays, port congestion, weather, or political developments, AI can flag potential disruptions early and recommend alternative routes or suppliers. Advanced systems even use digital twin simulations, creating virtual replicas of entire supply chains to test how they would respond under different crisis scenarios. The key benefit isn't just surviving shocks. The key benefit lies in being able to bounce back faster, sometimes even gaining competitive advantage while rivals' struggle.

Expanded Case Example: Maersk's AI-Driven Crisis Response

The Challenge

As the world's largest container shipping company, Maersk sits at the heart of global trade. During COVID-19 and subsequent global disruptions, Maersk faced massive challenges: port closures, unpredictable demand spikes, labor shortages, and congested shipping lanes. Traditional logistics planning could not keep pace with the sheer speed and complexity of disruptions.

The AI Solution

Maersk turned to AI-driven resilience tools to adapt in real time. Key initiatives included:

- Machine learning models to predict port congestion, container dwell times, and likely shipping delays.
- Dynamic routing algorithms to reroute vessels and containers around bottlenecks.
- Scenario simulations through digital twins, enabling Maersk to test responses to potential crises (e.g., sudden demand spikes or regional lockdowns).
- Predictive demand analytics for customers, helping them adjust orders and inventory before disruptions worsened.

The Results

- Maersk significantly reduced the impact of port bottlenecks by proactively rerouting shipments.
- Customers benefited from more accurate ETAs and greater transparency into supply chain risks.
- The company strengthened its reputation as a resilient logistics partner in an era of uncertainty.

- Lessons learned from COVID-19 informed long-term strategies for resilience against climate and geopolitical risks.

Lessons Learned

Maersk's experience underscores that resilience is not just about "absorbing" shocks. Resilience is also about anticipation and adaptation. By leveraging AI to predict and simulate disruptions, Maersk turned crises into opportunities to strengthen customer trust. The broader lesson: in an unpredictable world, resilience can be a real competitive advantage.

📌 Guiding Questions for Reflection

- How prepared is our supply chain for the next disruption — whether pandemic, conflict, or climate event?
- Do we rely too heavily on efficiency at the expense of resilience?
- Could AI-driven scenario simulations help us prepare for "what if" events before they strike?

Closing Thought

AI is quickly becoming the backbone of modern supply chains. Where traditional approaches focused on efficiency and cost-cutting, AI enables something far more powerful: the ability to anticipate, adapt, and respond in real time. From forecasting demand at the SKU level to rerouting shipments mid-journey, AI transforms supply chains from rigid systems into living networks that learn and evolve. The strategic advantage lies in the shift from reactive management to predictive and adaptive allocation. Instead of scrambling to fix problems after they occur — stockouts, delays, or supplier failures — AI gives businesses the foresight to spot issues early and the flexibility to act decisively. Companies that adopt this mindset are no longer victims of disruption; they become masters of agility.

Looking forward, AI-enabled supply chains will be true competitive differentiators. Customers will choose partners who can deliver reliably despite global shocks. Regulators and investors will favor firms with transparent, sustainable supply chains. And employees will thrive in organizations where resources are allocated intelligently and fairly. In short, AI isn't just optimizing the supply chain, it's redefining it. The businesses that embrace this transformation will set the standard for resilience, sustainability, and long-term growth in the decades to come.

Please refer to the below appendices located at the end of the book for chapter-specific resources. However, be aware that inclusion in the appendices should not be considered an endorsement by the author for any individual commercial product.

APPENDIX S: AI-Powered Demand Forecasting & Inventory Optimization Platforms

APPENDIX T: AI-Powered Logistics & Transportation Optimization Platforms

APPENDIX U: AI-Powered Resource Allocation & Workforce Scheduling Platforms

APPENDIX V: AI-Powered Supplier Risk Management Platforms

CHAPTER 11
AI IN CUSTOMER EXPERIENCE AND ENGAGEMENT

IN TODAY'S BUSINESS LANDSCAPE, products and prices can be copied in months, sometimes weeks. What can't be so easily duplicated is customer experience (CX). That's why companies across industries now see CX as their most important competitive battleground. According to PwC (2018), 73% of customers say experience is a key factor in their purchasing decisions. And many will pay more for a product if the experience is superior. The challenge is that delivering consistently excellent experiences is hard. Traditional CX strategies relied on human service teams, manual feedback collection, and generic loyalty programs. But these often fell short: long wait times frustrated customers, personalization was shallow at best, and companies struggled to anticipate problems before they escalated. This is where AI is transforming the game. By analyzing vast amounts of customer data in real time, AI makes it possible to deliver personalization at scale, anticipate customer needs before they're voiced, and provide service around the clock. In effect, AI allows companies to combine the empathy of human touch with the efficiency of automation.

Consider how customer expectations have shifted. Ten years ago, customers accepted waiting on hold for ten minutes. Today, they expect instant answers via chatbots, personalized recommendations in their shopping feeds, and seamless experiences across channels. Companies like Amazon, Netflix, and Apple have set the bar — and now customers expect the same level of responsiveness everywhere, from banks to airlines to retail stores. AI-powered CX tools address these challenges by:

- Analyzing behavior and preferences to deliver tailored product recommendations.
- Automating service with chatbots and virtual assistants that resolve issues instantly.
- Measuring sentiment across reviews, social media, and call transcripts to identify risks before customers' loss.
- Unifying data across touchpoints to create seamless omnichannel journeys.

The result isn't just happier customers — it's stronger business performance. Research from McKinsey shows that companies that excel in CX grow revenues 4–8% faster than their market peers (McKinsey & Company, 2019). In other words, CX is no longer just "nice to have." It's a driver of growth and loyalty.

AI gives businesses a powerful toolkit to meet these rising expectations. The companies that succeed won't be those that simply automate customer interactions, but those that use AI to create human-centered, trustworthy, and meaningful experiences.

📌 **Guiding Questions for Reflection**

- Do we treat customer experience as a cost to manage, or as a strategic asset to grow?
- Where are our biggest CX gaps today — speed, personalization, consistency, or empathy?
- How could AI help us deliver not just faster service, but more human-feeling engagement at scale?

Personalized Recommendations and Content

Personalization has always been the holy grail of customer experience. For decades, businesses have tried to make customers feel recognized and valued, with loyalty cards, birthday coupons, or segmented email lists. But these approaches were often too broad, lumping people into categories rather than truly tailoring experiences to the individual. AI has changed the game by enabling personalization at scale. Today, machine learning algorithms analyze browsing history, purchase patterns, demographics, and even real-time behavior to deliver highly tailored recommendations. Instead of "people like you bought this," companies can now predict what *you specifically* are most likely to want, sometimes before you even realize it yourself.

This ability to personalize recommendations and content is more than just convenience. Personalized recommendations drive measurable business impact:

- E-commerce companies using AI-powered recommendations see 10–30% increases in revenue (McKinsey & Company, 2021).
- Personalized product suggestions boost conversion rates and average order values.
- Customers reward personalization with loyalty. 80% say they're more likely to buy from a company that offers tailored experiences (Epsilon, 2018).

Personalization also creates an emotional effect: customers feel seen, valued, and understood. That sense of recognition builds trust, which in turn drives repeat business and lifetime value.

Expanded Case Example: Netflix's AI Recommendation Engine

The Challenge

Netflix's value proposition depends on keeping customers engaged. With tens of thousands of titles available, customers could easily feel overwhelmed. Traditional search and genre filters weren't enough to help people find content they loved. If users couldn't quickly find something to watch, they might cancel their subscriptions.

The AI Solution

Netflix built one of the world's most advanced recommendation engines, powered by AI. The system:

- Analyzes each user's viewing history, watch time, and even the points at which they pause or stop watching.
- Compares patterns across similar users to suggest likely interests.
- Continuously learns in real time as customers engage with the platform.
- Uses contextual data (time of day, device used, current trends) to refine recommendations.

Netflix even personalizes artwork and thumbnails for shows and movies, displaying different images based on what's most likely to attract each viewer's attention.

The Results

- Netflix estimates that 80% of the content streamed on its platform comes from personalized recommendations (Gómez-Uribe & Hunt, 2016).
- Engagement increased dramatically, reducing churn rates and keeping customers subscribed.
- The recommendation engine is considered one of Netflix's most important competitive advantages, fueling its global growth.

Lessons Learned

Netflix shows that personalization isn't just about driving sales, it's about curating meaningful experiences. By making content discovery effortless and engaging, Netflix turned personalization into the cornerstone of its customer strategy. The broader lesson: when personalization is done well, it feels less like marketing and more like service.

📌 Guiding Questions for Reflection

- Are we offering customers experiences that feel generic or truly personalized?
- Could AI recommendations improve not only sales, but also engagement and retention?
- How can we ensure personalization feels helpful rather than intrusive?

Conversational AI and Virtual Assistants

Customer service has always been one of the toughest parts of the customer experience to scale. Traditional call centers relied on human agents to answer every question, resolve every issue, and walk customers through even the simplest tasks. The result? Long wait times, high costs, and frustrated customers who felt like their time wasn't valued.

Enter conversational AI — chatbots and virtual assistants powered by natural language processing (NLP) and machine learning. Unlike old-fashioned automated phone menus ("Press 1 for billing…"), today's AI assistants can actually understand intent, carry on natural conversations, and solve problems instantly.

The benefits are twofold:

- For customers: 24/7 support, faster resolution, and answers without waiting on hold.
- For businesses: lower service costs, reduced call volumes, and the ability to free human agents to handle more complex, high-value interactions.

According to Juniper Research, AI-powered chatbots are expected to save businesses $8 billion annually by 2024, largely by automating routine interactions (Juniper, 2020). But the real value isn't just in cost savings, it's in customer satisfaction. When customers can resolve an issue in seconds instead of minutes, they remember it as a positive experience.

And importantly, the best implementations don't replace human agents, they augment them. Customers can escalate from bot to human seamlessly, ensuring that empathy and expertise are available when needed.

Expanded Case Example: Bank of America's Erica

The Challenge

We first introduced Erica, Bank of America's virtual assistant, back in Chapter 5 when we were discussing marketing and customer engagement. But this same example is applicable here as well. If you recall, Bank of America serves more than 67 million customers. Its call centers and branch offices were overwhelmed with routine requests: balance inquiries, bill payments, transaction history, and financial advice. These simple questions were eating up time and resources, leaving agents less able to focus on higher-value conversations. Customers wanted fast, digital-first answers — but without sacrificing trust and accuracy.

The AI Solution

Bank of America launched Erica in 2018. Erica is an AI-powered virtual financial assistant that has been integrated directly into Bank of America's mobile banking app. Erica uses natural language processing to understand spoken or typed requests, and connects to real-time financial data to deliver immediate answers.

Erica's capabilities include:

- Answering questions like *"What's my current balance?"* or *"Show me my recent transactions."*
- Proactively alerting customers to unusual spending or upcoming bills.
- Guiding users toward financial wellness with tips on saving and credit management.
- Seamlessly handing off to a human banker when queries become more complex.

The Results

- By 2022, Erica had engaged in over 1 billion interactions with customers.
- Adoption rates soared, with more than 25 million customers using Erica regularly (Bank of America, 2021).
- Customer satisfaction improved thanks to instant access to information and proactive alerts.
- Call center volumes declined, reducing operational costs while improving agent focus on complex cases.

Lessons Learned

Erica demonstrates how conversational AI can improve customer satisfaction and experience through automated personalization when providing customer support. By ensuring seamless human support when needed, Bank of America turned its virtual assistant into a core part of the customer experience without sacrificing customer trust. The lesson: conversational AI works best when it's not just reactive, but proactive and integrated into customers' daily lives.

🎯 **Guiding Questions for Reflection**

- How many of our customer interactions are still routine questions that could be automated?
- Could a conversational AI free up our human teams to deliver higher-value service?
- Are we designing our bots to be transactional — or truly conversational and proactive?

Sentiment Analysis and Voice of the Customer

If customer experience is the new battleground, then understanding the voice of the customer (VoC) is the intelligence that wins the war. Businesses have always gathered feedback through surveys and focus groups, but those methods are slow, limited in scope, and often biased toward the most vocal customers. Today, customers leave behind a vast trail of unstructured feedback: tweets, Instagram posts, online reviews, call transcripts, chatbot interactions, and more. Hidden in that data is a goldmine of insight, but only if businesses can analyze it at scale.

This is where AI-powered sentiment analysis comes in. By applying natural language processing (NLP) and machine learning, companies can gauge customer emotions, detect emerging trends, and even predict churn risks. AI doesn't just categorize comments as positive or negative, it can identify nuances like sarcasm, urgency, or frustration, and track how sentiment changes over time.

For businesses, the impact is significant:

- Early warning system: Spot product issues or PR crises before they escalate.
- Churn (loss) prediction: Identify at-risk customers based on negative sentiment patterns.
- Campaign optimization: Test how marketing messages are received in real time.

According to Deloitte, companies that leverage customer analytics are 23 times more likely to outperform competitors in customer acquisition and 9 times more likely to surpass them in customer loyalty (Deloitte, 2021).

Expanded Case Example: Coca-Cola's AI Social Sentiment Analysis

The Challenge

Here's another throwback to Chapter 5. As a global brand with billions of customers, Coca-Cola faced the challenge of keeping a pulse on consumer sentiment across markets. Traditional surveys were too slow to detect trends in real time. Social media, meanwhile, was exploding with feedback, both positive and negative, but the sheer volume made it impossible for human teams to track effectively. So around the same time they launched the "Create Real Magic" campaign using generative AI tools, Coca-Cola also began using AI-powered sentiment analysis platforms to analyze social media.

The AI Solution

Coca-Cola turned to AI-powered sentiment analysis platforms to monitor conversations across Twitter, Instagram, Facebook, and other channels. The system used NLP and machine learning to:

- Identify spikes in positive or negative mentions around campaigns, events, or product launches.
- Distinguish between casual mentions and urgent complaints.
- Analyze not only text but also images and emojis, detecting brand presence and associated emotions.
- Provide real-time dashboards to marketing and customer service teams.

The Results

- Coca-Cola could spot product issues within hours, not weeks, enabling rapid response.
- Marketing teams optimized campaigns mid-flight based on real-time feedback.
- Insights uncovered regional differences in consumer preferences, guiding local strategies.
- The company strengthened its reputation by engaging with customers more quickly and authentically.

Lessons Learned

Coca-Cola's case shows that AI sentiment analysis isn't just about listening, it's about acting fast and acting smart. By turning noisy, unstructured social chatter into structured insights, Coca-Cola was able to protect its brand, tailor experiences, and build stronger emotional connections. The broader lesson: companies that listen at scale — and respond authentically — earn long-term loyalty.

- How quickly can we detect shifts in customer sentiment today?
- Are we only hearing from the "loudest" voices, or capturing the full spectrum of feedback?
- How could real-time sentiment analysis help us prevent churn or reputational damage?

AI in Omnichannel Experiences

Modern customers don't just shop in one channel. They browse online, try products in-store, read reviews on social media, and then complete a purchase through a mobile app. This omnichannel reality means businesses must deliver seamless experiences across every touchpoint, a task that can be nearly impossible with siloed systems and manual processes. The challenge is consistency. Customers expect that when they interact with a brand, their history, preferences, and context travel with them. If a customer adds items to an online cart but later walks into a store, they don't want to start over. If they chat with a bot and then call customer service, they don't want to repeat their problem. Failing to integrate these touchpoints creates friction, which directly impacts loyalty.

AI helps close this gap by unifying data across channels and predicting where customers are in their journey. Machine learning can identify whether a customer browsing online is likely to convert in-store, or whether a social media inquiry signals churn risk. With this insight, companies can engage customers at the right time, through the right channel, with the right message. According to Accenture, companies with strong omnichannel strategies achieve 91% higher year-over-year customer retention rates compared to those without (Accenture, 2020). AI makes these strategies scalable, ensuring that personalization and consistency don't break down as customer journeys grow more complex.

Expanded Case Example: Sephora's AI-Powered Omnichannel Customer Journeys

The Challenge

Sephora, a global beauty retailer, faced the challenge of serving a digitally savvy customer base that expected personalized recommendations across online, mobile, and in-store experiences. Customers wanted to test products physically while still enjoying the ease of digital engagement. Siloed systems risked creating a disjointed experience.

The AI Solution

Sephora deployed a suite of AI-powered tools to bridge channels and deliver a unified experience:

- Virtual Artist app: Using augmented reality (AR) and machine learning, customers could "try on" makeup virtually via their phones.
- Personalized recommendations: AI engines suggested products based on past purchases, browsing behavior, and skin profiles.
- In-store integration: Beauty Advisors in physical stores accessed customer profiles generated by AI, providing consistent, personalized guidance.
- Chatbots and digital assistants: Handled common questions, linked directly to the customer's purchase history, and nudged customers with tailored offers.

The Results

- Increased engagement: Customers spent more time in both digital and physical channels.
- Higher conversion rates, as virtual try-on tools boosted purchase confidence.
- Stronger brand loyalty, with customers reporting a seamless experience between online browsing and in-store shopping.

- Valuable data insights, as Sephora gained a holistic view of customer preferences across touchpoints.

Lessons Learned

Sephora's case demonstrates that true omnichannel engagement isn't about adding more channels. The solution is to connect the ones you have intelligently. AI serves as the glue between the channels, ensuring customers experience the brand as a single, consistent journey rather than a series of fragmented interactions. The broader lesson: when omnichannel works, it transforms shopping from transactional to relational.

📌 **Guiding Questions for Reflection**

- Are our customers experiencing our brand as one seamless journey, or as disconnected touchpoints?
- How well do our systems share data across digital and physical channels?
- Could AI help us predict and guide customers across channels more effectively?

Proactive Customer Support

Customer service has traditionally been reactive: the customer has a problem, they call or email, and the company responds. But in today's hyper-competitive market, reactive service isn't enough. Customers expect companies to anticipate their needs and solve issues before they even notice them. This is where AI-powered proactive support comes into play. By analyzing real-time data, such as flight delays, system outages, subscription renewals, or product usage patterns, AI can predict when problems are likely to occur and trigger automatic solutions or outreach.

Instead of waiting for a frustrated call, companies can send a helpful message:

- *"Your package is delayed; here's a credit for the inconvenience."*

- *"We noticed your subscription is expiring; would you like to renew with one click?"*
- *"A flight disruption is expected; we've already rebooked you on the next available option."*

The impact is enormous: proactive support reduces churn, boosts satisfaction, and transforms customer relationships from transactional to trust-based. According to Gartner (2020), companies that successfully implement proactive service can increase customer retention by up to 15%

Expanded Case Example: Delta Airlines' AI-Powered Proactive Rebooking

The Challenge

For airlines, delays and cancellations are inevitable. Traditionally, passengers discovered disruptions only when they checked flight boards or received last-minute announcements at the gate. The burden was on them to queue at service desks or call hotlines, leading to frustration, long waits, and poor customer experiences.

The AI Solution

Delta Airlines deployed AI-driven tools that monitored flight schedules, weather patterns, and operational data in real time. When disruptions became likely, the system proactively:

- Identified affected passengers.
- Predicted the best alternative itineraries.
- Automatically rebooked passengers on the most suitable flights.
- Sent notifications directly via mobile apps, texts, or emails, often before passengers even reached the airport.

AI also personalized rebooking by factoring in traveler preferences, such as avoiding overnight layovers or maintaining seat class.

The Results

- Passengers avoided long lines at service counters and received solutions instantly.
- Customer satisfaction improved, with fewer negative experiences during disruptions.
- Delta strengthened loyalty by showing it valued passengers' time and comfort.
- Operational efficiency improved, as call centers and gate agents handled fewer crises.

Lessons Learned

Delta's approach shows that proactive support doesn't just minimize inconvenience — it creates moments of delight in situations that are traditionally frustrating. By reimagining service as something predictive, AI helped Delta turn disruptions into opportunities to build customer trust. The broader lesson: proactive AI support shifts companies from problem-based to experience-based.

📌 **Guiding Questions for Reflection**

- How many of our customer interactions are still reactive, waiting for complaints to come in?
- Could AI help us predict and resolve issues before customers notice them?
- What opportunities do we have to turn a traditionally negative moment into a positive surprise?

Ethical Considerations in AI Customer Experience

AI-powered customer experience is powerful. But with that power comes responsibility. While personalization, proactive support, and sentiment analysis can delight customers, they can also cross the line into creepy, unfair, or invasive if not handled carefully.

Three main ethical risks stand out:

1. Over-Personalization and the "Creepy Factor"

Personalization is valuable, but there's a fine line between helpful and unsettling. When customers feel like a company knows *too much* about them, or uses their data in ways that feel intrusive, trust erodes quickly. For instance, a well-known case involved Target's predictive analytics inadvertently revealing a teenager's pregnancy to her family based on her shopping patterns (Duhigg, 2012). The lesson: transparency matters. Customers should understand why they're receiving recommendations and have the ability to control or limit personalization.

2. Algorithmic Bias

AI systems are only as fair as the data they're trained on. If historical data reflects biased patterns, for example, disproportionately negative reviews of certain accents in call transcripts or unequal access to financial services, AI may perpetuate or even amplify those biases. This creates risks of discrimination in customer service, loan approvals, or pricing. Companies must adopt bias auditing and fairness checks to ensure AI doesn't unintentionally disadvantage certain groups.

3. Privacy and Data Protection

AI thrives on data, but that creates risks around privacy. Regulations like the General Data Protection Regulation (GDPR) in Europe and the California Consumer Privacy Act (CCPA) in the U.S. set strict rules on how customer data can be collected, stored, and used. Customers increasingly expect transparency and control. A Cisco survey found that 84% of consumers care about data privacy, and 48% have switched companies over privacy concerns (Cisco, 2022). If businesses misuse or fail to protect customer data, the reputational fallout can be severe.

Balancing AI and Ethics

To mitigate these risks, companies need guiding principles:

- Transparency: Explain why recommendations or decisions are made.

- Choice: Allow customers to opt out or adjust personalization.
- Fairness: Audit algorithms regularly to detect and correct bias.
- Security: Treat customer data as a valuable asset that must be safeguarded.

Handled correctly, AI can create trust by showing customers that personalization and automation are designed to serve them, not exploit them. Done poorly, it risks turning AI-powered CX into a liability instead of a strength.

📌 Guiding Questions for Reflection

- Do our personalization strategies enhance customer trust — or risk undermining it?
- Are we auditing our AI systems for fairness and bias?
- Do customers feel in control of their own data and choices?
- Are we using AI in customer experience to enhance relationships or primarily to cut costs?
- Do our personalization efforts feel helpful to customers — or do they risk crossing into the "creepy factor"?
- How quickly can we detect shifts in customer sentiment, and how fast do we act on them?
- Are our omnichannel experiences truly integrated and seamless, or do customers still feel like they're starting over at every touchpoint?
- Do we have adequate guardrails in place — transparency, fairness checks, and privacy protections — to ensure AI builds trust rather than erodes it?

Closing Thoughts

Customer experience has become one of the most important competitive differentiators in modern business. Customers no longer choose brands based only on product or price — they choose the companies that make their lives easier, more personalized, and more engaging.

AI is central to delivering on this expectation. From recommendation engines to conversational assistants, from sentiment analysis to omnichannel integration, AI makes it possible to provide human-centered experiences at scale. Companies like Netflix, Sephora, Coca-Cola, and Delta show that when executed well, AI can delight customers while also driving loyalty, revenue, and operational efficiency.

But the real lesson is this: AI is not a substitute for human connection. AI is an enabler of it. The businesses that succeed will be those that use AI not to replace empathy, but to amplify it. By combining AI's speed and scale with human judgment and authenticity, organizations can build experiences that feel both personal and trustworthy. Looking ahead, the companies that master AI in customer experience will not just win sales, they will win loyalty, trust, and lifetime value. And in a marketplace where expectations continue to rise, that is the ultimate competitive advantage.

CHAPTER 12

INDUSTRY-SPECIFIC AI APPLICATIONS

ARTIFICIAL INTELLIGENCE IS OFTEN DESCRIBED in sweeping, transformative terms. But in practice, AI's impact is highly industry-specific. A hospital, a retailer, and a factory floor face very different challenges, data environments, and regulatory constraints. For this reason, applying AI effectively requires understanding industry context.

For example:

- In healthcare, data is highly sensitive, tightly regulated under laws like HIPAA in the U.S. and GDPR in Europe, and often fragmented across electronic health record systems. AI applications must prioritize patient privacy and clinical accuracy above all.
- In retail, customer data is plentiful, but success depends on delivering personalization that feels helpful rather than invasive. Here, speed, recommendation accuracy, and omni-channel integration are key.
- In finance, AI adoption is shaped by strict compliance requirements. Algorithms must be explainable, auditable, and fair — especially in lending or risk assessments (European Banking Authority, 2020).

- In manufacturing, sensor data from industrial equipment is massive but often unstructured. The opportunity lies in predictive maintenance, supply chain optimization, and automation.

This chapter explores AI through the lens of five major industries — healthcare, retail, finance, real estate, and manufacturing — showing not only how AI is applied, but how data quality, regulation, and customer expectations influence its success.

🪁 Guiding Questions for Reflection

- In our industry, what unique constraints (e.g., regulations, legacy systems) shape AI adoption?
- Are we benchmarking against peers in our sector to understand competitive advantage?
- Do we risk applying "generic" AI tools without tailoring them to our industry context?

AI in Healthcare

Few industries generate as much data as healthcare. Every day, hospitals, clinics, and insurers produce mountains of information — from medical images and lab results to electronic health records (EHRs), prescription data, insurance claims, and even wearable device metrics. The irony is that while healthcare is *overflowing* with data, much of it sits in silos. Different systems won't or can't always talk to each other. A patient might have lab results stored in one hospital's database, scans in another, and insurance data in a third, with no easy way to connect the dots.

This is exactly where artificial intelligence begins to show its value. AI thrives on large, complex datasets. The more varied and abundant the data, the more insights AI can extract. That is, if the data is clean, structured, and integrated. In healthcare, that means turning raw data into tools that help clinicians make faster, more accurate diagnoses, optimize hospital operations, and even discover new drugs.

Think about medical imaging. Radiologists examine hundreds of scans daily, often under intense time pressure. Fatigue can lead to missed details. AI algorithms trained on millions of images can flag suspicious patterns, such as a shadow that might indicate lung cancer, or micro-calcifications that suggest early-stage breast cancer. The goal isn't to replace the radiologist, but to act as a second set of eyes, catching things a human might miss, or prompting a closer review.

Then there's the challenge of patient flow. Hospitals are constantly juggling admissions, discharges, and bed availability. AI-powered scheduling systems can predict peak times for ER visits, suggest optimal staffing levels, and even anticipate which patients are at higher risk of readmission. That kind of foresight translates into shorter wait times, smoother operations, and better patient outcomes.

AI is also reshaping drug discovery, traditionally a slow and expensive process. Pharmaceutical companies spend billions screening chemical compounds to find potential treatments. With AI, researchers can model how molecules interact with the human body, quickly narrowing millions of possibilities down to a handful of promising candidates. During the COVID-19 pandemic, AI systems were used to identify potential antivirals far faster than traditional methods would allow.

In short, healthcare has all the ingredients for AI to make a profound difference: vast datasets, critical decisions that depend on speed and accuracy, and huge stakes in terms of both lives and dollars. But the real promise isn't just efficiency — it's about creating a system where clinicians are empowered with better tools, patients receive faster diagnoses, and treatments get to market sooner. AI in healthcare isn't just a technological upgrade. It's about delivering better care, at scale, in a world where demand continues to rise.

Applications include:

- Medical imaging: Deep learning models detect tumors and anomalies in scans.
- Predictive diagnostics: Algorithms anticipate sepsis, heart failure, or readmission risks.

- Operational efficiency: AI optimizes patient flow, staff scheduling, and billing.
- Drug discovery: AI accelerates molecule screening and trial simulations.

Expanded Case Example: Google Health in Breast Cancer Detection

The Challenge

Breast cancer is one of the leading causes of death among women. Radiologists face heavy workloads, and mammogram interpretation has variability in accuracy. False negatives can delay treatment. False positives can cause unnecessary anxiety and procedures.

The AI Solution

Google Health developed an AI model trained on tens of thousands of mammograms. Using deep learning, the system could identify subtle patterns that humans might miss. Importantly, the model was designed to assist, not replace, radiologists.

The Results

- In trials, the AI system reduced false positives by 5.7% and false negatives by 9.4% compared to human radiologists alone (McKinney et al., 2020).
- When paired with radiologists, the combined human + AI workflow achieved even higher accuracy.

Lessons Learned

AI in healthcare is most effective as an augmenting tool. It reduces diagnostic errors, improves consistency, and helps overworked clinicians. However, human oversight remains critical. AI should be a partner, not a substitute, in clinical care.

📌 Guiding Questions for Reflection

- Which clinical or administrative areas in healthcare could AI safely improve?

- How do we balance innovation with strict regulatory compliance?
- What safeguards are needed to maintain trust in AI-assisted care?

AI in Retail and E-Commerce

Retail is one of the most data-saturated industries in the world. Every time a customer browses a website, clicks on a product, adds something to a cart, makes a purchase, or even abandons a checkout page, they're generating data. And it doesn't stop there. Loyalty programs track shopping habits, mobile apps monitor browsing behavior, and in-store sensors can capture foot traffic patterns. All of this adds up to a goldmine of information. But there's far too much for humans alone to process. This is where AI really shines. With its ability to analyze huge volumes of data in real time, AI helps retailers personalize, predict, and optimize customer experiences in ways that were once impossible.

Take personalization, for example. In the past, a retailer might segment customers broadly: "young professionals," "families," or "retirees." But today, AI enables one-to-one personalization. That means recognizing not just the segment a customer belongs to, but what *that specific individual* is most likely to want next. Think of the way Amazon suggests products you didn't even realize you were looking for. AI is working behind the scenes, drawing patterns from millions of customer journeys.

Prediction is another area where AI is transforming retail. Demand forecasting used to rely on spreadsheets and historical averages. Now, machine learning models can analyze seasonality, regional trends, weather forecasts, and even social media buzz to predict what customers will want and when. For a retailer, that means having the right inventory in the right place at the right time, minimizing stock-outs, reducing excess inventory, and boosting profit margins.

Optimization ties it all together. AI isn't just about what to stock or what to recommend; it's about streamlining the entire customer journey. That includes chatbots that answer questions instantly, dynamic pricing that adjusts to market conditions, and warehouse robots that speed up fulfillment. For customers, the experience feels seamless: personalized recommendations, faster delivery, and fewer frustrations. For retailers, it means higher sales, greater efficiency, and more loyalty.

The beauty of AI in retail is that it touches both front-end experiences and back-end operations. Shoppers see better product suggestions and smoother service, while companies quietly benefit from leaner supply chains and smarter pricing strategies. Done right, it creates a win-win: customers feel understood, and businesses perform better.

Applications include:

- Personalized recommendations and targeted promotions.
- Demand forecasting and inventory optimization.
- Automated customer support with chatbots.
- Dynamic pricing strategies.

Expanded Case Example: Amazon's Recommendation Engine

The Challenge

With millions of products, Amazon risked overwhelming customers with choice. Traditional browsing and search filters weren't sufficient to drive conversions at scale.

The AI Solution

Amazon built an AI-driven recommendation engine powered by collaborative filtering and deep learning. It analyzed:

- Customer browsing and purchase histories.
- Similar customer behaviors ("people who bought this also bought...").
- Real-time contextual signals (device, location, time of day).

The Results

- 35% of Amazon's revenue is attributed to its recommendation engine (McKinsey, 2013).
- Increased average order value by cross-selling and upselling.

- Reinforced customer loyalty by creating a sense of personalization at scale.

Lessons Learned

AI personalization drives not just revenue, but also customer constancy. The more customers engage, the more data the AI collects, the better the personalization becomes. This interaction creates a feedback loop that strengthens many aspects of the business over time.

📌 **Guiding Questions for Reflection**

- Are we using customer data to deliver personalization, or just generic promotions?
- How do we balance effective recommendations with avoiding "creepy" overreach?
- Could AI help us optimize supply chains alongside customer experience?

AI in Finance and Banking

Few industries have embraced data-driven decision-making as eagerly as finance. For decades, banks and investment firms have relied on models and algorithms to price risk, evaluate portfolios, and detect fraud. But what's happening now is a step change: AI is expanding what's possible beyond traditional analytics, offering speed, accuracy, and scale that humans alone can't match.

Think about fraud detection. In the past, banks would flag suspicious activity based on simple rules, such as a credit card being used in two different countries within an hour. But fraudsters have become more sophisticated, and simple rules no longer cut it. Today's AI-powered fraud detection systems can sift through millions of transactions in real time, spotting subtle anomalies, like unusual purchase patterns, abnormal device usage, or odd login behavior, enabling them to shut down fraud before it causes damage.

AI is also transforming customer experience. Instead of waiting in long phone queues, customers can now interact with intelligent chatbots or virtual assistants, available 24/7. These aren't the clunky bots of the past; they can answer detailed account questions, suggest financial products, and even provide financial wellness tips. Bank of America's *Erica*, for example, has become one of the most widely used digital assistants in banking.

Then there's investment and trading. Hedge funds and asset managers are increasingly turning to AI-driven models that can analyze massive datasets. They can analyze anything, from market prices and corporate filings to satellite imagery and social sentiment, all in record time, to spot opportunities humans might miss. While algorithmic trading isn't new, AI has made these systems far more adaptive, capable of learning and adjusting strategies dynamically as markets shift.

Even in the back office, AI is making a difference. Consider compliance and regulatory reporting. Financial institutions deal with mountains of documentation, contracts, disclosures, loan applications, and much of it in unstructured text. Natural language processing (NLP) tools can now read, extract, and classify information from these documents with remarkable accuracy, saving thousands of hours of manual review.

Of course, finance is also one of the most heavily regulated sectors. This means AI applications must meet high standards of explainability and accountability. A bank can't simply say, "The algorithm decided this person doesn't qualify for a loan." Regulators and customers expect to know *why* these decisions are made. They also expect customers personal data to be secure. This means financial institutions adopting AI must balance innovation with transparency and security, ensuring models are explainable, fair, and absolutely secure.

In short, AI in finance and banking is about more than efficiency — it's about building smarter, faster, and safer systems that protect consumers, optimize investments, and reduce the administrative burden. When done right, it creates a triple win: for customers (better service), for institutions (greater efficiency), and for regulators (more consistent compliance).

Applications include:

- Fraud detection through anomaly detection.
- Algorithmic trading and portfolio optimization.
- Customer service chatbots.
- Automated contract review and compliance checks.

Expanded Case Example: JPMorgan's COiN Platform

The Challenge

JPMorgan processes over 12,000 commercial credit agreements annually. Manual contract review was costly, time-consuming, and prone to errors.

The AI Solution

The bank launched the Contract Intelligence (COiN) platform, which uses NLP to analyze legal documents and extract key clauses in seconds.

The Results

- Tasks that took lawyers 360,000 hours annually were reduced to seconds (JPMorgan, 2017).
- Improved accuracy in detecting risks and compliance issues.
- Freed legal teams to focus on high-value advisory work.

Lessons Learned

AI can dramatically reduce manual drudgery in compliance-heavy sectors. But success depends on ensuring models are auditable and explainable — critical in finance, where regulators demand accountability.

📌 Guiding Questions for Reflection

- What repetitive compliance or documentation tasks could AI handle?
- How do we ensure transparency and fairness in AI-driven lending or trading?
- Could AI give us a competitive edge in detecting fraud earlier?

AI in Real Estate and Property Management

Real estate may not be the first industry people think of when it comes to cutting-edge AI, but it's undergoing a quiet transformation. Buying, selling, and managing property has traditionally been slow, paper-heavy, and based on human judgment. Now, AI is reshaping how homes are valued, how investments are made, and even how buildings are run day to day.

One of the most obvious uses is in property valuation. In the past, valuations relied on comparable sales, local expertise, and an appraiser's judgment. While valuable, this process was subjective and sometimes inconsistent. Today, AI models can analyze thousands of variables at once, such as recent sales, property characteristics, neighborhood crime rates, school quality, and even satellite imagery, to generate valuations that are faster and, in many cases, more accurate. AI is also enhancing the buying and renting experience. Virtual tours powered by AI and augmented reality let prospective buyers "walk through" a property from their laptop or phone. Natural language processing makes it easier for customers to search listings in plain English, for example asking for "a two-bedroom apartment with lots of light near a park" and receiving meaningful results. These tools make real estate more accessible and customer-friendly, especially for younger, tech-savvy buyers.

On the management side, AI is revolutionizing how buildings are operated. Smart property management platforms can monitor energy use, water systems, HVAC, and even security cameras to optimize efficiency and safety. Predictive maintenance, a familiar concept in manufacturing, is now being applied to building maintenance. AI can now anticipate when an elevator or air conditioning unit is likely to fail, enabling managers to act before tenants complain. Investors are also benefiting from predictive analytics. AI can scan regional housing trends, population growth, and economic signals to forecast property appreciation or rental yields. For large investment firms, this means smarter acquisition strategies. For smaller landlords, it means reducing risk in property decisions.

Of course, real estate also raises sensitive questions about fairness and ethics. If AI-driven valuation tools or tenant screening systems inherit bias from historical data, they risk reinforcing inequalities in housing access. That's why transparency and fairness matter just as much in this sector as in finance or healthcare.

In short, AI is modernizing a sector long resistant to change. From valuations and virtual tours to predictive property management, AI is making real estate transactions more efficient, transparent, and data-driven.

Applications include:

- Predictive analytics for property values and rental trends.
- AI-driven virtual tours and AR/VR staging.
- Smart building energy management.
- Tenant sentiment analysis from reviews and service requests.

Expanded Case Example: Zillow's Zestimate

The Challenge

For years, Zillow's "Zestimate" tool was a go-to feature for anyone curious about the value of their home. The problem? Early versions were often criticized for being inaccurate, sometimes off by tens of thousands of dollars. Homeowners grew frustrated, real estate agents distrusted it, and buyers were skeptical of relying on it to guide decisions. In a market where trust is everything, unreliable estimates risked undermining Zillow's brand.

The AI Solution

Zillow doubled down on its Zestimate, moving from basic statistical models to advanced machine learning and deep learning. The updated system pulled in far more data sources, including:

- Public property records and tax assessments.
- Historical sales and listing data.
- Property details like square footage, age, and renovation history.
- Satellite imagery and neighborhood features.
- Broader market trends like seasonality and regional demand.

The AI model wasn't just crunching numbers; it was learning relationships between variables. For example, it could weigh how proximity to good schools, recent kitchen upgrades, or even local traffic patterns affected value. By feeding the model millions of records, Zillow taught it to recognize complex patterns that human appraisers might miss.

The Results

The revamped Zestimate made a significant leap in accuracy. By 2019, Zillow reported a median error rate of less than 2% for on-market homes, a massive improvement over earlier versions (Zillow, 2019). This boosted user trust and kept customers engaged with the platform. In fact, many buyers and sellers started using Zestimate as a first step before contacting agents — cementing Zillow's role as a data-driven authority in the real estate space.

Lessons Learned

The Zestimate story highlights both the potential and the pitfalls of AI in real estate. On one hand, machine learning can dramatically improve accuracy, speed, and scalability compared to traditional methods. On the other, transparency is critical. Zillow had to clearly communicate that Zestimate was a starting point, not an official appraisal. By setting realistic expectations and continually improving the model, Zillow turned an early liability into a competitive advantage.

The broader takeaway? In real estate, where trust and transparency are paramount, AI must not only deliver better results but also explain itself in a way customers can understand.

📌 Guiding Questions for Reflection

- How can AI valuation tools shape buyer and seller expectations?
- Do predictive models risk reinforcing inequalities in housing access?
- Could AI-enabled smart building management improve both cost savings and sustainability?

AI in Manufacturing and Logistics

If retail is about delighting customers and finance is about managing risk, then manufacturing and logistics are about keeping the world running. From factories producing the goods we use every day to the supply chains that deliver them, efficiency and reliability are everything. This is exactly where AI is making some of the biggest, though often least visible, impacts.

Manufacturers have long relied on data, from quality checks to production scheduling, but AI takes this to another level. With thousands of IoT sensors embedded in machines, assembly lines now generate real-time streams of data throughput, such as temperature readings, vibrations, and power consumption. On its own, this data can feel overwhelming. But with AI, companies can detect subtle signals that predict when a machine is likely to fail, allowing maintenance teams to fix issues before they cause costly downtime. This approach, known as predictive maintenance, saves millions by replacing "fix it after it breaks" with "fix it before it fails."

AI also strengthens quality control. Instead of relying solely on human inspectors or random sampling, computer vision systems can analyze every product coming off the line. They catch defects invisible to the naked eye, ensuring higher consistency without slowing production. For industries like automotive, aerospace, or pharmaceuticals, where a single defect can have serious consequences, this ability is game-changing.

Logistics, the other half of this story, benefits just as much. Shipping networks are incredibly complex, balancing delivery times, fuel costs, and route planning. AI can optimize these variables in real time, rerouting trucks around traffic jams, predicting weather disruptions, or adjusting warehouse workflows based on incoming demand. The result? Packages arrive faster, fleets operate more efficiently, and costs are reduced without sacrificing reliability.

Perhaps most importantly, AI in manufacturing and logistics helps companies become more resilient. The COVID-19 pandemic exposed just how fragile global supply chains can be. Companies that had already invested in AI were better positioned to adjust quickly, using demand forecasting tools to anticipate shortages and AI-driven allocation systems to prioritize limited resources.

In short, AI is turning factories and supply chains into smart, adaptive systems. Instead of reacting to problems after they happen, businesses can anticipate issues, optimize workflows, and continuously improve. While customers may never see these behind-the-scenes changes directly, they certainly feel them when shelves are stocked, products arrive on time, and quality remains high.

Applications include:

- Predictive maintenance of equipment.
- Quality control with computer vision.
- Supply chain optimization and demand sensing.
- Robotics and automation on the factory floor.

Expanded Case Example: Siemens' Smart Factories

The Challenge

Manufacturing is a high-stakes environment. A single machine breakdown on a production line can halt output for hours, sometimes days, costing millions in lost productivity. Traditional maintenance approaches, like scheduled servicing or reactive fixes, often fell short. Either machines were serviced too early (wasting time and resources) or too late (leading to unplanned downtime). Siemens, a global leader in industrial automation, faced this exact challenge across its factories and for its clients. The company needed a smarter way to keep machines running, improve quality control, and boost efficiency without increasing costs.

The AI Solution

Siemens turned to artificial intelligence to build what it calls "smart factories." Using thousands of IoT sensors embedded across machines and assembly lines, Siemens collected continuous streams of data on temperature, vibration, power consumption, and more. Then, AI systems analyzed this data in real time to:

- Detect early warning signs of equipment failure and trigger predictive maintenance.
- Use computer vision to automatically inspect products, catching microscopic defects before they left the factory.
- Optimize production scheduling by predicting bottlenecks and adjusting workflows dynamically.

AI wasn't just bolted onto the process, it became the backbone of Siemens' "Digital Industries" strategy, enabling a holistic view of factory operations.

The Results

The shift paid off in measurable ways:

- Predictive maintenance cut unplanned downtime significantly, saving millions of euros in lost production.
- Automated quality control increased inspection accuracy while freeing human workers from repetitive tasks.
- Overall productivity in Siemens' Amberg factory improved to the point where 99% of product quality issues were identified before leaving the line (Siemens, 2021).

- Clients adopting Siemens' AI-powered solutions reported similar benefits, from reduced operating costs to faster time-to-market.

Lessons Learned

Siemens' experience shows that AI in manufacturing isn't about replacing workers with machines, it's about creating smarter, more resilient systems. The key lesson is that success depends on data. Without high-quality IoT sensor data, predictive models can't function. Another takeaway: AI works best when humans and machines collaborate. While algorithms monitor patterns and optimize schedules, human workers bring the problem-solving and adaptability needed on the factory floor.

In short, Siemens' smart factories prove that when AI is paired with rich sensor data and thoughtful integration, it can transform manufacturing from reactive and rigid to predictive and adaptive.

🎯 Guiding Questions for Reflection

- Could predictive maintenance cut our equipment downtime?
- How do we prepare workers to collaborate with AI-driven machines?
- Could supply chain AI help us anticipate disruptions more effectively?

Cross-Industry Lessons Learned

When you step back from the details of healthcare scans, retail recommendations, bank fraud detection, property valuations, and smart factories, a bigger picture starts to emerge. Despite operating in very different contexts, these industries reveal common patterns about what makes AI succeed, and what pitfalls to avoid.

1. Data Quality Is the Foundation Everywhere

No matter the sector, the story always begins with data. In healthcare, fragmented records limit what AI can achieve. In retail, personalized recommendations are only as good as the accuracy of purchase history.

In real estate, valuation models depend on clean property and neighborhood data. And in manufacturing, IoT sensors must generate reliable streams of information for predictive maintenance to work. The shared lesson? AI magnifies the quality of data it's given — good data drives good outcomes, but bad data spreads errors faster than ever.

2. AI Works Best as an Augmenter, Not a Replacement

Across industries, AI proves most effective when it supports human expertise rather than replacing it. For example, radiologists use AI as an additional set of eyes rather than a substitute for their skills. Bank lawyers utilize contract analysis tools but still apply their own judgment. Factory managers rely on predictive analytics when making decisions about workflows, but apply adaptable solutions to human problems. The consistent finding is that people plus AI achieve better results than either could on their own.

3. Transparency Builds Trust

Trust is the universal currency. In finance, regulators and customers demand clear explanations of why an algorithm denied a loan. In real estate, Zillow had to communicate that its Zestimate is a guide, not gospel. In healthcare, patients and clinicians alike want to understand how AI models arrive at diagnostic insights. Without transparency, even the most accurate model may be rejected. Trust isn't just nice to have, it's essential for adoption.

4. Ethical Risks Repeat Themselves

Bias, fairness, and privacy aren't industry-specific challenges; they're cross-industry risks. Whether it's a hiring algorithm in retail, a diagnostic tool in healthcare, or a valuation model in housing, AI can reinforce systemic inequalities if not carefully designed and monitored. Similarly, data privacy concerns, from GDPR in Europe to HIPAA in healthcare, surface everywhere. The lesson? Ethical guardrails must travel with AI wherever it goes.

5. AI Creates Feedback Loops That Grow Value Over Time

One striking pattern is how AI can get smarter the more it's used. Amazon's recommendations improve with each click and purchase.

Predictive maintenance systems get better as they process more sensor data. Healthcare diagnostic models become sharper as more scans are fed into training sets. This compounding effect means early adopters can gain a lasting competitive edge because their systems literally learn faster than latecomers'.

6. Human Readiness Matters as Much as Tech Readiness

Finally, there's culture. AI doesn't just require clean data; it requires people willing to use it, trust it, and adapt alongside it. Whether it's radiologists adjusting to new workflows, bankers trusting AI-driven fraud alerts, or property managers using smart systems to manage buildings, the common denominator is human adoption. Culture and training often determine whether AI becomes a game-changer or a wasted investment.

📌 Guiding Questions for Reflection

- Are we investing as much in data quality as in AI tools?
- How can AI complement our people, rather than try to replace them?
- Do we have enough transparency in our AI systems to build trust?
- What ethical safeguards do we need to prevent bias or misuse?
- Are we ready to embrace the feedback loop advantage that comes with early adoption?
- How do sector-specific regulations shape what's possible for us?

Closing Thought

AI is not a monolith. Its applications differ across healthcare, retail, finance, real estate, and manufacturing, but the underlying principles remain constant. The industries that thrive are those that tailor AI to their unique context while embracing best practices in data, culture, and governance. In the long run, AI will not just optimize processes within industries. AI will redefine industry boundaries, creating new models of healthcare delivery, retail engagement, financial inclusion, property management, and industrial production. The future belongs to organizations that see AI not only as a universal enabler, but as a tool to reimagine their industry's possibilities.

Please refer to the below appendices located at the end of the book for chapter-specific resources. However, be aware that inclusion in the appendices should not be considered an endorsement by the author for any individual commercial product.

APPENDIX X: Some of the Most Commonly Used AI Tools in Healthcare

APPENDIX Y: Some of the Most Commonly Used AI Tools in Retail and E-Commerce

APPENDIX Z: Some of the Most Commonly Used AI Tools in Real Estate and Property Management

APPENDIX AA: Some of the Most Commonly Used AI Tools in Manufacturing and Logistics

CHAPTER 13
BUILDING AN AI ROADMAP FOR YOUR BUSINESS

EXECUTIVE OVERVIEW: **What an AI Roadmap Must Do**

Every business leader today is under pressure to "do something with AI." But without a clear roadmap, AI efforts can quickly become scattered pilot projects that generate more confusion than value. An effective AI roadmap is not about chasing the latest tools — it's about creating a step-by-step plan that connects AI investments directly to business outcomes.

At its core, an AI roadmap must accomplish three things:

1. Anchor AI in Business Strategy

AI is not a side project for the IT team. It should be woven into the fabric of the company's growth, cost, and risk priorities. Whether the goal is to improve customer retention, cut supply chain costs, or enhance forecasting accuracy, the roadmap should make explicit how AI initiatives support core business objectives. Leaders should be able to answer: *"If this AI project succeeds, how does it move the needle on our strategy?"*

2. Balance Quick Wins with Long-Term Foundations

The most successful roadmaps don't try to "boil the ocean." They start small — identifying quick wins that demonstrate value in 90 days or less — while simultaneously laying the foundations for scale. That means building the data pipelines, governance structures, and cultural readiness needed for sustainable AI adoption. In practice, this looks like running a pilot chatbot to improve customer service while also investing in the company's data governance program.

3. Treat AI as a Portfolio, Not a One-Off

Just like financial investments, AI initiatives should be managed as a portfolio. Some projects will deliver immediate ROI, others will take longer but build strategic advantage, and a few may fail — and that's acceptable if the portfolio as a whole delivers value. A good roadmap provides mechanisms for prioritizing, reviewing, and, when necessary, stopping projects that aren't delivering. This disciplined approach reduces hype and ensures AI remains a measurable business asset, not an experimental playground.

Ultimately, the job of an AI roadmap is to make AI a normalized, well-managed part of the business plan. By anchoring AI in strategy, balancing wins with foundations, and treating initiatives as a portfolio, leaders can shift from dabbling with AI to systematically unlocking its potential at scale.

Readiness Assessment (Where You Are Now)

Before any business can chart a roadmap for AI, it needs to take a hard look at its starting point. Too often, organizations jump straight into pilots without first understanding their readiness. This can lead to stalled projects, wasted budgets, and skepticism about AI's value. A readiness assessment provides the baseline for realistic planning and helps leaders decide whether to walk, jog, or sprint into AI.

A strong readiness assessment looks at four key dimensions:

1. Strategy Fit

- Are we clear on our top three business priorities where AI could have impact (e.g., revenue growth, margin improvement, risk reduction)?
- Do we have leadership alignment on why we're pursuing AI?

This isn't about listing 50 potential use cases. It's about zeroing in on where AI can most directly support the company's goals.

2. Data Readiness

- Do we know where our critical business data resides?
- Is our data clean, consistent, and accessible? If not, how do we remedy this?

AI models are only as strong as the data that feeds them. A company with siloed, poor-quality data will struggle to deliver meaningful results no matter how advanced the algorithm.

3. People and Culture

- Do we have executive sponsorship for AI initiatives?
- Are our employees sufficiently data-literate to effectively use AI tools in their daily work?
- Is there an appetite for change, or resistance to "machines replacing humans"?

Culture is often the hidden barrier. If staff don't trust or understand AI, adoption will lag even if the technology works perfectly.

4. Technology and Process Infrastructure

- Do we have IT platforms that can support experimentation, deployment, and scaling?
- Are there existing Machine Learning Operations (MLOps) or Development Operations (DevOps) practices to manage AI models over time?
- Do we have the right vendor ecosystem, or will we need to build in-house?

Tech readiness doesn't mean having the most expensive tools. It means having an environment where AI pilots can run safely and scale if successful.

The output of this assessment is often summarized in a simple AI Readiness Heatmap: green for areas of strength, amber for areas needing investment, and red for critical gaps. This visual makes it clear where the organization stands and what gaps must be closed before scaling.

By pausing to assess readiness, companies avoid overpromising and underdelivering. Just as a construction team checks the strength of a foundation before building a skyscraper, business leaders should ensure their strategy, data, people, and technology are ready to carry the weight of AI.

Example: AI Readiness Heatmap

Dimension	Key Questions	Status	Notes
Strategy Fit	Are top business priorities defined and aligned with AI opportunities?	Green	Clear alignment on customer retention, cost reduction, and risk control.
Data Readiness	Is data clean, accessible, and integrated across systems?	Amber	CRM data is solid, but supply chain and finance data remain siloed.
People & Culture	Do leaders sponsor AI? Are employees data-literate and open to adoption?	Red	Pockets of enthusiasm, but low data literacy; fear of "AI replacing jobs."
Technology & Process	Do we have platforms, MLOps, and vendor support for scaling?	Amber	Cloud platform in place, but no formal MLOps or model governance.

Interpretation

- **Green (Strength):** The business strategy is well aligned with AI, making prioritization easier.
- **Amber (Needs Work):** Data and technology foundations exist but need investment before large-scale AI.
- **Red (Critical Gap):** Culture and skills represent the biggest barrier — without change management and training, adoption will stall.

This simple format makes it easy for executives to see at a glance where to invest first before scaling AI.

Use-Case Discovery (Where Value Lives)

Once you've assessed readiness, the next step is to identify where AI can create tangible business value. This is often where companies get stuck — either generating a laundry list of ideas that lack focus, or chasing trendy applications that don't tie back to strategy. The goal of use-case discovery is to find opportunities that are both impactful and realistic, given your current data, culture, and technology.

Start with Business Pain Points, Not Technology

The best use cases don't start with, *"We want to try generative AI."* They start with, *"We're losing customers because our service response is too slow"* or *"We're sitting on excess inventory."* By framing problems in terms of business outcomes — revenue lift, cost reduction, risk control, customer satisfaction — you ensure AI is being applied to what matters most for your organization, and not investing in fluff that won't help your bottom line.

Patterns to Look For

Across industries, most AI use cases fall into a few repeatable patterns:

- **Prediction** (e.g., forecasting demand, predicting equipment failure, anticipating churn).
- **Personalization** (e.g., tailoring offers, customizing learning pathways, dynamic recommendations).
- **Optimization** (e.g., supply chain routing, workforce scheduling, pricing models).
- **Generation** (e.g., creating content, drafting reports, simulating designs).
- **Anomaly Detection** (e.g., fraud detection, cybersecurity alerts, quality control).

Leaders can use these categories as a lens for brainstorming relevant opportunities in their own organizations.

Practical Methods for Discovery

1. **Customer Journey Mapping:** Walk through your end-to-end customer experience and ask: *Where are the biggest friction points? Can AI could improve outcomes?*
2. **Value Stream Mapping:** Chart your key business processes (e.g., order-to-cash, procure-to-pay) and look for bottlenecks.
3. **P&L Levers:** Analyze your profit and loss statement. Where does a 1–2% improvement in cost or revenue create the most impact?
4. **Voice of Employees:** Gather insights from frontline teams. They know the repetitive, manual tasks that are ripe for automation.

Examples in Practice

- **Retail:** Instead of saying, *"Let's try AI for marketing,"* focus on a clear pain point: *"Our promotions aren't driving conversion because they're not personalized."* AI-powered recommendation engines can directly target this issue.
- **Manufacturing:** Rather than starting with *"Let's deploy IoT sensors,"* reframe it as: *"Downtime is costing us millions per year."* Predictive maintenance becomes the high-value use case.
- **Finance:** Instead of *"We want an AI chatbot,"* frame the need as: *"Our call centers are overloaded, and response times are driving complaints."* AI assistants now become a business-backed solution, not a tech experiment.

The Use-Case Backlog

The output of this stage should be a living backlog of potential AI use cases. Each item should include:

- A short title.
- The business problem it addresses.
- The data required.
- A potential business owner.

- A rough sense of value and feasibility.

This backlog is not a commitment. It is simply a starting point for prioritization. Over time, ideas can move up, down, or off the list as you refine your understanding of AI opportunity vs business needs.

By treating use-case discovery as a disciplined, business-first exercise, leaders avoid the trap of chasing hype and instead build a pipeline of initiatives that deliver measurable impact.

📌 Guiding Questions for Leaders

- What business pain points or opportunities matter most in the next 12–18 months?
- Which processes are most prone to errors, delays, or inefficiencies?
- If we could only solve one customer friction point with AI, what would it be?

Prioritization Framework (What to Do First)

By now, you've likely generated a backlog of potential AI use cases. Some will be game-changers, some modest improvements, and some "nice to have" experiments. The challenge is that most organizations don't have unlimited resources, so the next step is to decide:

Which use cases should we pursue first?

A prioritization framework helps leaders cut through the noise and select the opportunities most likely to deliver value, quickly and safely.

Key Scoring Dimensions

When ranking use cases, consider evaluating them along five dimensions:

1. **Business Value**
   - How much impact could this use case have on revenue, margin, risk, or customer experience?
   - Is the benefit incremental or transformational?

2. **Feasibility**
   - Do we have the data, skills, and technology required?
   - How complex is the implementation?
3. **Risk & Compliance**
   - Are there regulatory or ethical risks (bias, privacy, fairness)?
   - Could it create reputational risk if it fails?
4. **Strategic Fit**
   - Does this initiative align with our company's strategic goals?
   - Will it strengthen competitive advantage?
5. **Time-to-Impact**
   - Can we show value in weeks or months, rather than years?
   - Does this lend itself to a pilot/MVP approach?

Each dimension can be scored on a simple 1–5 scale (1 = low, 5 = high). Adding up the scores provides a quick view of overall priority.

Example Use-Case Prioritization Scorecard

(Rate each use case on a 1–5 scale; add notes)

Use Case	Value (Impact on P&L, CX, or Risk)	Feasibility (Data, Skills, Tech)	Risk/Compliance Concerns	Strategic Fit	Time-to-Impact	Total Score	Notes
Example: Predictive Maintenance	5	4	2	5	4	20	Data from IoT sensors available; need pilot funding

Tools for Prioritization

1. **2×2 Matrix**
   - Plot Value (high vs. low) against Feasibility (high vs. low).
   - Focus on "high value, high feasibility" use cases first.

2. **Weighted Scorecard**
   - Assign weights to each dimension (e.g., Value 40%, Feasibility 25%, Risk 15%, Strategic Fit 10%, Time-to-Impact 10%).
   - Multiply scores by weights to get a composite score.
3. **MoSCoW Method**
   - Classify each use case as Must-have, Should-have, Could-have, Won't-have (for now).
   - This keeps teams realistic about capacity and focus.

Example: Prioritization Weighted Scorecard

Use Case	Value (1–5)	Feasibility (1–5)	Risk/Compliance (1–5)	Strategic Fit (1–5)	Time-to-Impact (1–5)	Total
Predictive Maintenance	5	4	3	5	4	21
AI Chatbot for Customer CX	4	5	4	4	5	22
Dynamic Pricing Model	5	3	2	5	3	18
AI Contract Review	3	4	5	3	4	19

Interpretation: The chatbot and predictive maintenance score highest overall — they are strong candidates for early pilots. The dynamic pricing model has high value but lower feasibility and higher risk, suggesting it should be a later-phase initiative once foundations are stronger.

Key Takeaway

Prioritization is not just about numbers, it's about making transparent, defensible choices. A structured framework helps executives avoid gut-feel decisions, align stakeholders, and ensure resources go toward the initiatives most likely to succeed.

📌 Guiding Questions for Leaders

- Which use cases have the potential to deliver visible wins in the next 3–6 months?
- Are we considering both value *and* feasibility, or just chasing the biggest idea?
- How do we balance "safe bets" with a few strategic moonshots?

Business Case & Funding Model

Once a use case has been prioritized, the next step is to justify the investment in terms leaders understand: dollars, risk, and outcomes. A business case proposal provides the bridge between technical promise and executive approval and financing. A strong AI business case proposal should be concise, pragmatic, and hypothesis-driven. It shouldn't try to predict every detail, but instead lays out the potential benefits, costs, and risks clearly enough to justify a pilot.

Key Elements of a One-Page Business Case Proposal:

1. **Use Case Title & Owner**
   - Who is sponsoring this initiative, and who is accountable for results?
   - Example: *"AI Chatbot for Customer Service — Sponsored by VP of Customer Experience."*
2. **Problem Statement**
   - What pain point or opportunity does this address?
   - Example: *"Average customer response time is 18 minutes, leading to 20% drop in satisfaction."*
3. **Proposed AI Solution**
   - Brief description of the model or system.
   - Clarify if the plan is to build, buy, or partner.
   - Example: *"Deploy an AI-powered virtual assistant to handle FAQs and triage service tickets."*
4. **Expected Benefits (Quantified if Possible)**
   - Revenue lift: ___%
   - Cost savings: ___%

- Risk reduction: ___%
- Example: *"Reduce call volume by 30%, saving $1.2M annually in contact center costs."*

5. **Costs & Resources Required**
   - Data prep, platform or license costs, staffing or vendor support.
   - Example: *"$150K platform license, 3 FTEs for implementation, $50K for data cleaning."*

6. **KPIs & Success Metrics**
   - What will success look like?
   - Example: *"Customer wait times reduced to under 5 minutes; 80%+ of routine inquiries resolved by chatbot."*

7. **Risks & Guardrails**
   - Compliance, fairness, customer trust, or technical risks.
   - Example: *"Ensure chatbot handoffs to human agents remain seamless; avoid biased responses."*

8. **Decision Recommendation**
   - Proceed / Revise / Park.
   - Framed as: *"Should we invest in a pilot of this use case now?"*

Funding Model Options

AI initiatives often require a mix of funding approaches:

- **Central Innovation Fund**: A dedicated pool for pilots and early experimentation, reducing the burden on individual business units.
- **Business Unit Funding**: Each department funds AI use cases that align with their budgets and KPIs.
- **Co-Funding Model**: Costs shared between central AI teams and business units, ensuring accountability.

The choice depends on maturity: early-stage organizations may centralize funding to encourage experimentation, while mature organizations push more ownership to specific business units.

Example Snapshot: Business Case for AI Chatbot

Title & Owner: AI Chatbot for Customer Service — VP, Customer Experience

Problem Statement: Long response times (18 min avg) reducing Net Promoter Score (NPS - a customer satisfaction metric) by 20%.

Solution: Deploy AI chatbot for FAQs and ticket triage (partner with a SaaS vendor).

Expected Benefits: Reduce call volume by 30% ($1.2M annual savings), improve NPS +10 points.

Costs: $150K vendor license, $50K data prep, 3 FTEs.

KPIs: Wait times <5 min, 80% resolution rate, NPS +10.

Risks/Guardrails: Bias testing, seamless human handoff.

Decision: Proceed to 90-day pilot.

📌 **Guiding Questions for Leaders**

- Can we articulate the problem and benefits in plain business terms?
- Do costs and risks align with expected impact?
- Are KPIs measurable within a 90-day pilot?

Build vs. Buy vs. Partner

Once you have a strong business case, the next question is: *How should we execute this?* Should we build the solution internally, buy a pre-built product, or partner with an external provider? The right answer depends on your organization's goals, resources, and risk appetite.

Option 1: Build (In-House Development)

Best when:

- The use case is strategic IP that differentiates you from competitors.

- You have, or want to build, internal AI talent (data scientists, ML engineers, MLOps).
- You need customization beyond what off-the-shelf tools can provide.

Example: A global retailer developing its own proprietary demand-forecasting engine, using unique transaction data to outperform generic models.

Risks: Higher upfront cost, longer time-to-market, risk of talent shortages.

Option 2: Buy (Off-the-Shelf Solutions)

Best when:

- The capability has been commercially developed and is widely available.
- Speed-to-value matters more than uniqueness.
- Vendors already meet compliance, security, or industry standards.

Example: A mid-size bank deploying a SaaS chatbot platform for customer service instead of building its own NLP model.

Risks: Limited customization, ongoing license costs, potential vendor lock-in.

Option 3: Partner (Hybrid or Vendor Collaboration)

Best when:

- You need speed and customization.
- You lack in-house expertise but want to learn and co-develop.
- The use case benefits from specialized domain vendors (e.g., healthcare AI providers, logistics optimizers).

Example: A manufacturer partnering with Siemens to deploy predictive maintenance, combining vendor algorithms with in-house sensor data.

Risks: Dependence on vendor roadmap, shared control of IP, integration complexity.

Decision Factors

When weighing Build vs. Buy vs. Partner, ask:

1. Differentiation — Does this capability set our business apart, or is it simply the minimum required to allow us to be competitive in our market?
2. Time-to-Value — How fast do we need results?
3. Talent Availability — Do we have (or can we hire) the skills needed to build? To use effectively if we buy? If neither are true, partnership may be a necessity.
4. Total Cost of Ownership (TCO) — Beyond licenses, what will we spend on data optimization, integration, maintenance, and upgrades?
5. Risk & Compliance — Does one option reduce regulatory or security risk more than others?
6. Vendor Lock-In — Are we comfortable being tied to a vendor's roadmap and pricing model?

Rules of Thumb

- Buy when the capability is standardized (optical character recognition, generic chatbots, cloud analytics).
- Build when the capability is core IP that drives your unique advantage (pricing engines, proprietary customer models).
- Partner when the capability requires specialization or speed, but your business needs to learn in order to succeed or eventually own more of the process.

📌 **Guiding Questions for Leaders**

- Does this use case differentiate us or just keep us competitive?
- How fast do we need impact — weeks, months, or years?
- What risks come with outsourcing critical parts of our business intelligence?

Pilot Design ("Prove It Fast, Safely")

Even with a solid business case and a clear build/buy/partner decision, the smartest way to begin is with a **pilot**. Pilots give organizations the chance to validate assumptions, test risks, and learn quickly without overcommitting resources. Done well, they create momentum. Done poorly, they can sour executives and staff on AI altogether.

The principle is simple: prove it fast, but prove it safely.

Key Components of a Strong AI Pilot

1. **Clear Scope**
   - Define exactly what the pilot will test — and what it won't.
   - Example: A customer service chatbot pilot might handle only FAQs and password resets, not complex billing issues.
2. **Time-Bound**
   - Set a timeline (usually 6–12 weeks) with clear start and end dates.
   - Avoid pilots that drag on indefinitely and drain credibility.
3. **Success Metrics & Kill/Scale Thresholds**
   - Define upfront what success looks like.
   - Example KPIs: accuracy above 85%, cost savings of $50K per quarter, or reduction in customer wait times by 30%.
   - If metrics aren't met, pause or "kill" the pilot; if exceeded, prepare to scale.
4. **Data Pipeline Plan**
   - Identify the data needed, how it will be collected, and how it will be cleaned.
   - Example: For a predictive maintenance pilot, ensure sensor data is consistently available and reliable.
5. **Human-in-the-Loop Workflow**
   - Build in oversight so humans validate or override AI outputs during testing.
   - Example: Fraud detection models flag suspicious transactions, but analysts review them before action is taken.

6. **Risk Controls**
    - Conduct bias testing, privacy checks, and security reviews before launch.
    - Document how exceptions or failures will be handled.

Sample Pilot Design Snapshot:

Use Case: AI Chatbot for Customer Service

Scope: Handle FAQs and password resets only

Duration: 10 weeks

Success Metrics:

- 80% accuracy on routine queries
- Reduction in call volume by 25%
- Customer satisfaction ≥ 85%

Data Needs: Historical chat logs and FAQs integrated into chatbot training

Human Oversight: All escalated queries routed to live agents; agent feedback loop improves training data

Kill/Scale Threshold: If accuracy <70% or customer satisfaction drops, stop pilot. If metrics are met or exceeded, plan for enterprise rollout.

Why Pilots Fail (and How to Avoid It)

- **Too vague:** No clear scope or objectives, leading to endless testing.
- **Too ambitious:** Trying to solve everything in one pilot.
- **No success criteria:** Teams declare victory regardless of results.
- **No exit strategy:** Pilots that never end, wasting time and money.

The most successful pilots are laser-focused, time-bound, and outcome-driven — building both confidence and clarity for the next stage. Having a Pilot Charter helps to keep these objectives clear from the start.

✒ Guiding Questions for Leaders

- Have we clearly defined what's in scope and out of scope for this pilot?
- What metrics will tell us to stop, revise, or scale?
- How are we ensuring oversight, safety, and customer trust during testing?

Platform Foundations (Scale Later Without Rework)

Pilots are exciting because they produce quick wins, but many companies fall into the "pilot graveyard" trap, dozens of small successes that never scale. The common reason? Weak data and IT platform foundations. If the pilot can't plug into enterprise systems or if the data pipeline isn't repeatable, the project stalls after its first success. To avoid this, leaders need to invest early in the IT architecture, governance, and processes that will allow AI to move from experiment to production smoothly.

Core Data Foundations

1. **Data Architecture**
   - Decide how data will be stored and accessed (e.g., warehouse vs. lakehouse).
   - Standardize data storage formats across systems so pilots don't rely on "one-off" integrations.
2. **Integration**
   - Ensure the technical tools and methods used to connect different software systems and allow the exchange of information between them, typically known as Application Programming Interfaces (API) and event streams, connect core systems (Enterprise Resource Planning software, Customer Relationship Management systems, supply chain management software, etc.).

- Example: An AI model predicting inventory needs must pull from both sales and supplier data, not just one data silo. Therefore, these data silos need to be connected.

3. **Data Quality and Governance**
   - Implement regular audits, cleaning routines, and master data management procedures.
   - Define ownership: who is accountable for the accuracy of which data sets?

4. **Security & Privacy Controls**
   - Learn the compliance regulations that apply to your organizations data collection and usage, and assure all regulatory standards can be met, before implementing any new changes.
   - Apply access controls and audit logging early, so pilots don't have to be "retrofitted" for compliance later.

Platform Foundations for AI

1. **MLOps Practices**
   - Treat models like products: version them, track performance, monitor for drift.
   - Build pipelines for CI/CD (continuous integration/continuous deployment) of AI models.

2. **Scalability**
   - Choose platforms that can handle both today's pilot and tomorrow's enterprise load.
   - Cloud-native solutions are often the most flexible for scaling.

3. **Monitoring & Feedback Loops**
   - Set up dashboards to monitor accuracy, fairness, and usage.
   - Example: A loss prediction model should trigger alerts when accuracy drops below a predetermined threshold.

Why This Matters

Without these foundations, pilots get "stuck." A chatbot that worked in one business unit can't be rolled out globally because it doesn't integrate with other systems. A demand forecasting tool delivers insights, but executives don't trust the numbers because data governance is weak. By contrast, companies that lay data and platform foundations early can scale faster, reduce duplication, and build trust across the organization.

Case Snapshot: Starbucks

Starbucks invested heavily in its Deep Brew AI platform, ensuring data from loyalty apps, point-of-sale systems, and supply chain logistics flowed into a unified system. Because of this integration, the same AI infrastructure powers personalized marketing, store staffing optimization, and supply planning. Instead of dozens of isolated pilots, Starbucks created a platform for continuous innovation.

📌 Guiding Questions for Leaders

- Do we have a clear data architecture that avoids silos?
- Are our pilots designed with security, governance, and compliance in mind from day one?
- Can today's pilot scale across business units, or will it need to be rebuilt later?

Operating Model & Governance

Even with the right data and platform foundations, AI won't scale without the right operating model and governance framework. In many organizations, AI initiatives stall because they are scattered across business units, lack clear ownership, or run into regulatory roadblocks. A strong operating model ensures AI is not just a set of experiments but a repeatable, well-managed business capability.

Operating Model Options

There is no one-size-fits-all model, but three common structures have emerged:

1. **Center of Excellence (CoE)**
   - A central AI team builds expertise, develops best practices, and drives enterprise-wide standards.

- Best for organizations just starting, to avoid duplication and ensure consistency.
- Risk: Can become a bottleneck if too centralized.

2. **Hub-and-Spoke**
    - A central hub defines standards and platforms, while business units (spokes) develop and deploy AI tailored to their needs.
    - Balances consistency with flexibility.
    - Works well for mid-size to large organizations scaling across multiple functions.

3. **Embedded Squads**
    - AI talent is embedded directly in business units, working alongside domain experts.
    - High proximity to problems and users, which accelerates adoption.
    - Risk: Can lead to fragmentation without central oversight.

Most mature organizations evolve toward a hybrid model, usually central governance with decentralized innovation.

Defining Roles & Responsibilities

A clear RACI (Responsible, Accountable, Consulted, Informed) helps avoid confusion:

- **Product Owner:** Defines business goals, prioritizes use cases.
- **Data Engineer:** Prepares and pipelines data for AI models.
- **ML Engineer:** Builds, trains, and deploys models.
- **Compliance Officer:** Ensures regulatory, ethical, and risk standards are met.
- **Business Lead:** Champions adoption within the function.
- **IT/Infrastructure:** Manages integration, security, and scalability.

Without clarity, projects drift, accountability blurs, and adoption suffers. Developing a RACI Matrix is one way to stay on top of the specific roles and responsibilities for any AI project.

Example RACI Matrix (Roles & Responsibilities)

Activity	Product Owner	Data Engineer	ML Engineer	Compliance	Business Lead	IT/Infra	Notes
Define use case	R	C	C	C	A	I	
Collect/clean data	C	A	R	I	C	I	
Build/train model	I	C	A/R	I	C	C	
Test/validate	C	R	A	C	C	I	
Deploy model	I	C	A	I	R	A	
Monitor/report	C	R	A	C	I	C	

R = Responsible, A = Accountable, C = Consulted, I = Informed

Governance Essentials

AI governance is about ensuring that models are safe, fair, explainable, and well-managed. Common practices include:

1. **Model Risk Committee**
   - Reviews high-impact AI models before deployment.
   - Similar to credit risk committees in finance.
2. **Ethics Review Board**
   - Evaluates fairness, bias, and transparency issues.
   - Ensures alignment with company values and societal norms.
3. **Change Advisory Process**
   - Monitors model updates, ensuring new versions are tested and approved.
   - Prevents "shadow AI" from sneaking into production.
4. **Documentation Standards**
   - Model cards: Summaries of purpose, data sources, performance, and limitations.
   - Audit logs: Who changed what, when, and why.

Case Snapshot: Microsoft's Responsible AI Governance

Microsoft created a Responsible AI Standard with guiding principles and governance boards across ethics, legal, and engineering. By formalizing its operating model, Microsoft avoided fragmented adoption and built customer trust by embedding fairness and transparency directly into its AI practices.

📌 Guiding Questions for Leaders

- Do we have a clear operating model for AI — centralized, hybrid, or decentralized?
- Who is accountable for ensuring AI projects meet ethical, legal, and compliance standards?
- Are our governance practices lightweight enough to enable innovation but strong enough to prevent misuse?

Change Management & Adoption

The toughest part of AI transformation isn't the technology — it's the people. Even the best-designed models and platforms will fail if employees don't understand them, trust them, or see how they fit into daily work. That's why change management and adoption are as critical as algorithms and data pipelines.

Why Adoption Fails

Many AI projects stall after pilots, not because the model underperformed, but because:

- Employees weren't trained to use the tool.
- Teams felt threatened by automation, fearing job loss.
- Leaders failed to communicate why the change mattered.
- Incentives rewarded old behaviors rather than adoption of new ones.

Without a deliberate adoption strategy, AI can feel like something being *"done to people"* rather than *"done with them."*

Winning Buy-In

1. **Narrative from the Top**
   - Leaders must frame AI not as cost-cutting automation but as a growth and empowerment tool.
   - Example: Instead of *"AI will replace call center jobs,"* say *"AI will free agents from repetitive queries so they can focus on higher-value conversations."*
2. **Quick Wins, Widely Shared**
   - Celebrate early pilot successes visibly to build momentum.
   - Example: *Share testimonials from frontline employees whose work improved because of AI.*
3. **Involve Users Early**
   - Engage employees in testing prototypes and giving feedback.
   - Co-creation builds trust and ownership.

Training & Capability Building

- **Executive Education:** Leaders need to understand AI's opportunities, risks, and limitations in plain language.
- **Role-Specific Training:** Provide tailored learning paths, including data literacy for all employees, advanced training for technical teams, and applied workshops for managers.
- **"Day-in-the-Life" Playbooks:** Show employees how AI fits into their workflows, not just in theory but in practice.

Incentives & Performance Metrics

- Align performance goals with AI adoption.
- Example: *Sales teams could be measured not just on revenue but also on effective use of AI-assisted lead scoring.*
- Incentives should reward outcomes achieved with AI, not just old manual methods.

Communication & Engagement Cadence

- **Regular Updates:** Town halls, newsletters, and dashboards to show progress.
- **Office Hours & Champions:** Create a network of AI champions across functions who answer questions about model usage.
- **Feedback Loops:** Build mechanisms for employees to raise concerns, suggest improvements, and share success stories.

Case Snapshot: AT&T's Workforce Reskilling

AT&T invested over $1 billion in its reskilling program to prepare employees for AI-driven roles. By combining learning platforms with clear communication about career pathways, the company not only closed skill gaps but also boosted morale. Employees felt AI was part of their career growth, not a threat to it.

📌 Guiding Questions for Leaders

- Are we telling a clear, inspiring story about *why* AI matters for our people?
- Do employees have the training and playbooks they need to succeed with AI?
- Are incentives aligned to reward adoption rather than cling to old methods?

Scaling What Works (From Pilot to Platform)

Pilots prove whether AI can deliver results, but scaling determines whether those results create lasting competitive advantage. Many organizations end up stuck in the "pilot purgatory" stage of their AI project, with dozens of successful small projects that never translate into enterprise transformation. To escape this, leaders need a disciplined process for scaling what works.

When to Scale

Not every pilot should grow. Scale only when:

- **Sustained ROI** is demonstrated over time, not just in one-off results.
- **Metrics are stable,** with no major drift or unexplained anomalies.
- **Users adopt and trust** the system consistently.
- **Risks are manageable,** with compliance and security controls in place.

If these conditions aren't met, refine or kill the pilot rather than wasting resources.

How to Scale Successfully

1. **Harden the Infrastructure**
   - Move from experimental data pipelines to production-grade systems.
   - Implement platform improvements to ensure uptime, reliability, and monitoring standards across the organization.
   - Example: *A forecasting model that worked in spreadsheets must be migrated into a cloud platform integrated with Enterprise Resource Planning software.*
2. **Standardize the Capability**
   - Treat AI not as one-off projects but as reusable building blocks.
   - Example: *Instead of building a new recommendation engine for each business unit, design one shared capability with adjustable parameters.*
3. **Expand Across Functions**
   - Identify adjacent processes that can benefit from the same capability.
   - Example: *A predictive maintenance model for one plant can be adapted across multiple factories with minor tweaks.*

4. **Portfolio Thinking**
    - Manage AI initiatives like a portfolio: scale winners, retire underperformers, and continuously invest in new bets.
    - Balance quick wins with long-term differentiators (sustainable competitive advantages unique to your business).
5. **Industrialize Governance**
    - Embed review processes, bias testing, and audit logs as standard. Ensure Standard Operating Procedures are reviewed by the Subject Matter Experts before, and periodically after, implementation.
    - Scaling without governance risks inconsistent results and regulatory exposure.

Case Snapshot: Walmart's AI Forecasting at Scale

The Challenge: Walmart's manual forecasting struggled to keep pace with millions of SKUs and rapidly shifting consumer demand. Early AI pilots showed promise but weren't consistent across regions.

The AI Solution: Walmart invested in a unified AI platform for demand forecasting, integrating supply chain, store, and e-commerce data.

Results: The system reduced stockouts, improved accuracy by double digits, and allowed faster decision-making across thousands of stores.

Lesson Learned: Scaling required moving from fragmented pilots to a single, standardized platform accessible across the enterprise.

Why Scaling Matters

Scaling converts AI from *isolated wins* into enterprise muscle. It allows the same algorithms, data, and governance frameworks to generate value repeatedly, across functions and geographies. Without scaling, AI remains a collection of interesting projects. With scaling, it becomes a core capability of the business.

✒ Guiding Questions for Leaders

- Which of our AI pilots have shown sustained ROI and adoption?
- How can we design reusable capabilities rather than one-off models?
- Do we have governance in place to scale safely across the enterprise?

Time-Phased Roadmap (0–30–90–180–365 Days)

One of the biggest challenges leaders face with AI is knowing how fast to move. Go too slow, and competitors leap ahead. Move too fast without foundations, and projects collapse under their own weight. A time-phased roadmap balances speed with discipline, helping organizations build momentum while reducing risk. The timeline below is a practical blueprint — not rigid dates, but a structured sequence that ensures quick wins while laying the groundwork for sustainable scale.

0–30 Days: Establish the Baseline

- Conduct an AI readiness assessment (strategy, data, people, technology).
- Build an initial use-case backlog, capturing problems and opportunities across the business.
- Align leadership around top 3–5 strategic priorities for AI.
- Define governance guardrails (ethics, compliance, privacy).

☞ **Outcome:** *A clear snapshot of where your organization stands and where to focus.*

31–90 Days: Launch Quick Wins

- Select 1–2 high-priority use cases with strong business value and high feasibility.
- Develop pilot charters with scope, KPIs, and success thresholds.

- Stand up a cross-functional team (business + tech + compliance).
- Begin data remediation for critical gaps (cleaning, integration).
- Start executive and frontline AI awareness training.

☞ **Outcome:** *Early pilots in motion, with visible sponsorship and momentum.*

91–180 Days: Prove and Prepare for Scale

- Run pilots to completion (6–12 weeks each), measuring against predefined KPIs.
- Conduct kill/scale reviews to decide which pilots graduate.
- Harden data and platform foundations — scalable pipelines, monitoring, MLOps practices.
- Document playbooks and lessons learned from early pilots.
- Establish a central or hybrid AI operating model (CoE, hub-and-spoke).

☞ **Outcome:** *First wins validated, technical foundations upgraded, governance formalized.*

181–365 Days: Scale and Institutionalize

- Scale the highest-performing pilots into enterprise capabilities.
- Expand training and upskilling programs for broader adoption.
- Integrate AI KPIs into business scorecards and OKRs.
- Mature governance committees (ethics board, model risk review).
- Start a rolling portfolio process: continuously discover, prioritize, and scale new use cases.
- Communicate results widely to build confidence and momentum.

☞ **Outcome:** *AI moves from isolated projects to a strategic, repeatable business capability.*

Key Takeaway

This roadmap shows that meaningful AI adoption doesn't require five years of investment before value appears. By structuring progress into 30–90–180–365-day phases, leaders can deliver quick wins, build trust, and create the foundations for enterprise-wide transformation — without overwhelming the organization.

📌 Guiding Questions for Leaders

- Do we have a clear 30-day plan to align leadership and assess readiness?
- Which quick wins will we prioritize in the first 90 days?
- By day 365, what capabilities should be standard operating practice, not pilots?

Metrics & Scorecards

A roadmap without measurement is just wishful thinking. To ensure AI delivers real business value, leaders need metrics and scorecards that track outcomes, adoption, and governance. These indicators make AI initiatives transparent and accountable, preventing hype from over-shadowing results.

Four Categories of AI Metrics

1. **Business Outcome KPIs**
   - Measure the impact on P&L, customers, and operations.
   - Examples:
     - Revenue lift (e.g., +5% from personalized recommendations).
     - Cost savings (e.g., $1M annual reduction in call center costs).
     - Risk reduction (e.g., fewer fraud incidents, improved forecast accuracy).
     - Customer experience (e.g., Net Promoter Score ↑, churn ↓).

2. **Model Performance KPIs**
   - Measure how well the AI itself is working.
   - Examples:
     - Accuracy, precision, recall, F1 score.
     - Calibration (how well predicted probabilities match reality).
     - Drift detection (is performance degrading over time?).
3. **Adoption & Usage KPIs**
   - Measure whether people are actually using the system.
   - Examples:
     - % of employees using AI-assisted tools weekly.
     - Task completion times before vs. after AI.
     - Frontline satisfaction with AI-enabled workflows.
4. **Governance & Risk KPIs**
   - Ensure AI is safe, fair, and compliant.
   - Examples:
     - Number of models reviewed by ethics/risk committees.
     - Audit pass rates for compliance checks.
     - Bias/fairness test results across demographics.
     - Number of incidents flagged and resolved.

Scorecard Example:

Category	KPI	Target	Current	Status
Business Outcome	Customer wait times ↓	30% reduction	22%	● Amber
Business Outcome	Annual cost savings	$1.2M	$900K	● Amber
Model Performance	Chatbot accuracy	≥85%	88%	● Green
Adoption	% of service agents using chatbot	75%+	50%	● Red
Governance	Bias check compliance	100%	100%	● Green
Governance	Incidents escalated	<5 per quarter	2	● Green

This type of scorecard gives leaders a balanced view: not just *"is the model accurate?"* but also *"is it delivering value, being used, and staying compliant?"*

Why This Matters

AI success is multidimensional. A model with 95% accuracy that no one uses is a failure. A tool that saves costs but introduces bias is a reputational risk. Balanced scorecards force leaders to look at the full picture — value, adoption, performance, and ethics.

📌 **Guiding Questions for Leaders**

- Do we measure AI success in business terms, or just technical accuracy?
- Are adoption and governance KPIs given as much weight as ROI?
- Do we review metrics regularly enough to spot drift, bias, or declining adoption?

Risk, Ethics, and Compliance (Always-On)

AI may drive innovation, but without safeguards it can expose organizations to legal, ethical, and reputational risks. A roadmap that ignores governance is incomplete. Risk, ethics, and compliance need to be embedded from the start, not added on at the end. The mindset shift is this: AI risk management is not a project milestone, it's an *"always-on"* discipline.

Key Risk Domains to Manage

1. **Bias & Fairness**
   - AI trained on skewed data can amplify discrimination in hiring, lending, or customer service.
   - Example: An AI hiring system that favors resumes with certain demographics because of biased historical data.
2. **Transparency & Explainability**
   - Black-box models erode trust. Business leaders and regulators increasingly demand *why* a model made a decision.
   - Example: In finance, regulators require explainable lending decisions under the EU AI Act and U.S. Equal Credit Opportunity Act.
3. **Privacy & Data Protection**
   - Data misuse can trigger lawsuits and fines under GDPR, CCPA, and similar laws.

- Example: Using customer data for AI personalization without consent can result in multimillion-dollar penalties.

4. **Security & Resilience**
   - AI systems can be targets for adversarial attacks, data poisoning, or breaches.
   - Example: Fraud detection systems may be manipulated if attackers understand how alerts are triggered.

5. **Accountability**
   - Clear lines of responsibility must exist. Who owns the outcomes of an AI model: the data science team, IT, or business leaders?

Compliance Anchors

- **GDPR (General Data Protection Regulation, EU)**: Requires lawful basis for data processing, consent management, and data minimization.
- **CCPA (California Consumer Privacy Act, US)**: Grants consumers rights to know, delete, and opt out of data sharing.
- **EU AI Act (Draft/Upcoming)**: Categorizes AI by risk level, with strict obligations for high-risk use cases (e.g., healthcare, finance, HR).
- **FTC Guidelines (US)**: Warn businesses against deceptive or unfair use of AI, emphasizing transparency and fairness.

Safeguards to Embed in the Roadmap

1. **Bias Testing as Default**
   - Every model should undergo fairness testing across demographic groups.
2. **Model Cards & Documentation**
   - Standard summaries of purpose, data sources, assumptions, and known limitations.
3. **Ethics Review Boards**
   - Multidisciplinary committees to review high-impact models before deployment.

4. **Audit & Monitoring**
   - Continuous monitoring for drift, anomalies, and unintended outcomes.
   - Regulatory audits built into regular operations, not treated as emergencies.
5. **Incident Response Playbooks**
   - Define how the organization will respond if an AI system causes harm (e.g., biased outcomes, false fraud flags).

Why *"Always-On"* Matters

AI doesn't stay static — models learn, data shifts, regulations evolve. Without ongoing governance, risks compound. By treating ethics and compliance as living systems, businesses not only avoid fines but also build trust with customers, employees, and regulators.

🎯 Guiding Questions for Leaders

Building an AI roadmap isn't just a technical exercise. It's a leadership challenge. The following questions are designed to help executives stress-test their plans before moving forward:

Strategic Alignment

- Which 2–3 use cases would most move our P&L or customer outcomes in the next 12–18 months?
- Are we clear on *why* we are pursuing AI — growth, efficiency, risk reduction, or all of the above?
- Do our AI initiatives align with our broader business strategy and competitive positioning?

Data & Foundations

- What data debt blocks our top use cases, and how quickly can we fix it?
- Do we have a scalable data architecture and governance model to avoid silos?
- Have we embedded privacy, bias testing, and security into our data strategy from day one?

Execution & Ownership

- Do we have clear owners and sponsors for each prioritized use case?
- Have we set success metrics and kill/scale thresholds for pilots?
- Where must we be world-class (build) vs. where can we be good enough (buy/partner)?

Culture & Adoption

- Do employees understand how AI supports them rather than replaces them?
- Have we created training pathways and day-in-the-life playbooks for adoption?
- Are incentives aligned to reward usage of AI systems, not just old methods?

Scaling & Governance

- Are our AI projects designed as reusable capabilities, not one-off experiments?
- Do we have governance structures (e.g., ethics board, risk committees) in place to scale responsibly?
- How will we monitor for drift, bias, and compliance once AI is deployed at scale?

Risk & Trust

- Do we have a plan for incident response if an AI system fails or causes harm?
- How will we ensure transparency and explainability in high-stakes decisions?
- What steps are we taking to build trust with customers, regulators, and employees?

Key Insight:

If leaders can answer these questions with clarity and confidence, they are well-positioned to move from AI pilots to enterprise transformation. If not, the roadmap needs refining before execution begins.

Tools & Templates

An AI roadmap is easier to execute when leaders have structured tools at their disposal. This section provides a reference set of templates and checklists that can used as is or tailored to suit specific needs. Each tool is designed to be simple, repeatable, and adaptable across industries.

1. Use-Case Prioritization Scorecard

- Purpose: Rank potential AI use cases by value, feasibility, risk, fit, and time-to-impact.
- Format: Table with 1–5 scoring for each dimension, plus a total score.
- Use: Populate during discovery workshops to compare opportunities side by side.

2. One-Page Business Case Template

- Purpose: Turn a prioritized use case into a concise investment justification.
- Sections: Problem statement, proposed AI solution, expected benefits, costs, KPIs, risks, decision.
- Use: Executive-ready document for steering committees or funding reviews.

3. Pilot Charter Template

- Purpose: Ensure each pilot is scoped, time-bound, and measurable.
- Sections: Title, owner/team, scope, duration, objectives, data requirements, human oversight, kill/scale thresholds.
- Use: Prevents *"pilot purgatory"* by setting clear success/failure criteria upfront.

4. RACI Matrix (Roles & Responsibilities)

- Purpose: Clarify who is Responsible, Accountable, Consulted, and Informed for AI projects.
- Categories: Product owner, data engineer, ML engineer, compliance officer, business lead, IT/infra.
- Use: Helps avoid confusion and turf wars, especially in cross-functional teams.

5. KPI Scorecard Template

- Purpose: Track AI progress across business outcomes, model performance, adoption, and governance.
- Example: *"Customer wait times reduced by 30%," "Chatbot accuracy ≥85%," "Bias checks passed."*
- Use: Balanced view for executive dashboards, ensuring projects are delivering value *and* staying compliant.

6. AI Readiness Heatmap

- Purpose: Visualize current maturity across strategy, data, people, and technology.
- Format: Table or heatmap (Green = strength, Amber = needs work, Red = critical gap).
- Use: Early diagnostic tool to identify where to invest before scaling.

7. Risk & Governance Checklist

- Purpose: Standardize bias testing, privacy, security, and compliance checks for every model.
- Sections: Data sourcing, model documentation, fairness testing, audit logs, incident response.
- Use: Integrate into governance committee reviews before deployment.

How to Use This Toolkit

- **Workshop Mode:** Use the prioritization scorecard and backlog to identify opportunities.
- **Executive Review:** Summarize with the business case template and readiness heatmap.
- **Pilot Launch:** Apply the pilot charter and RACI matrix for execution.
- **Scaling Phase:** Monitor with the KPI scorecard and risk checklist.

By embedding these templates into the AI roadmap process, leaders create consistency, transparency, and discipline, three qualities that prevent pilots from stalling and enable AI to scale responsibly.

Closing Thought

AI is often portrayed as futuristic, disruptive, and almost magical. But for organizations seeking to harness it effectively, the real power of AI lies in making it a normal, well-managed part of business strategy rather than a flashy experiment. An effective AI roadmap achieves exactly that. It transforms AI from hype into habit by:

- Grounding initiatives in business priorities so every project ties directly to revenue growth, cost savings, or risk management.
- Balancing quick wins with long-term foundations so momentum builds without sacrificing scalability.
- Embedding governance and ethics from the start so AI is trusted, compliant, and sustainable.
- Scaling what works so the enterprise avoids "pilot purgatory" and builds lasting capability.
- Keeping people at the center by focusing on adoption, training, and trust.

The organizations that succeed are not those chasing the shiniest algorithms, but those that treat AI with the same discipline as finance, operations, or supply chain. They see it as a portfolio to be managed, a capability to be scaled, and a responsibility to be governed.

By following the roadmap, leaders can move beyond scattered pilots and position AI as a systematic enabler of business strategy. The companies that make AI a dependable business function will be the ones that quietly and powerfully redefine their industries.

Please refer to the below appendices located at the end of the book for chapter-specific resources/templates to assist in creating your unique AI Roadmap are included in:

APPENDIX 1: AI Roadmap Templates

APPENDIX 1: AI ROADMAP TEMPLATES

1. Use-Case Prioritization Scorecard

Rank potential AI use cases by value, feasibility, risk, fit and time-to-impact. Use this table to compare opportunities side by side.

Use Case	Value (1-5)	Feasibility (1-5)	Risk/Compliance (1-5)	Strategic Fit (1-5)	Time-to-Impact (1-5)	Total Score	Notes
Example: Predictive Maintenance	5	4	3	5	4	21	High potential, some risk

2. One-Page Business Case Template

Use Case Title: _____

Owner/Champion: _____

Problem Statement:

- What pain point or opportunity are we solving?
- Why now?

Proposed AI Solution:

- Short description of model/system.
- Build, buy, or partner?

Expected Benefits:

- Revenue lift: _____
- Cost savings: _____

- Risk reduction: _____

Costs/Resources Required:

- Data prep: _____
- Platform/licensing: _____
- People/partners: _____

KPIs to Track:

- _____

Risks & Guardrails:

- Compliance?
- Bias?
- Data privacy?

Decision: Proceed / Revise / Park

3. Pilot Charter Template

Pilot Title: _____

Owner/Team: _____

Timeline: Start _____ / End _____

Scope (What's In/Out):

- In: _____
- Out: _____

Objectives & Success Metrics

- Metric 1: Target _____
- Metric 2: Target _____
- Kill/Scale Threshold: _____

Data Requirements

- Sources: _____
- Cleaning/Labelling Needs: _____

Human-in-the-Loop Plan

- Who reviews AI outputs?
- How are overrides logged?

Risk Controls

- Bias testing: Y/N
- Privacy by design: Y/N
- Security controls: _____

Review Date: _____

4. RACI Matrix (Roles & Responsibilities)

Activity	Product Owner	Data Engineer	ML Engineer	Compliance	Business Lead	IT/Infra	Notes
Define use case	R	C	C	C	A	I	
Collect/clean data	C	A	R	I	C	I	
Build/train model	I	C	A/R	I	C	C	
Test/validate	C	R	A	C	C	I	
Deploy model	I	C	A	I	R	A	
Monitor/report	C	R	A	C	I	C	

R = Responsible, A = Accountable, C = Consulted, I = Informed

5. KPI Scorecard Template

Use Case	Business Outcome KPIs	Model Performance KPIs	Adoption KPIs	Governance KPIs	Current Status	Notes
Example: AI Chatbot	Avg handle time ↓20%	Precision 85%, Recall 90%	500 daily active users	Bias check: pass	☑ On Track	Next review in 30 days

AI Plan Readiness Checklist:

Strategic Alignment

- Are we clear on which 2–3 use cases would most move our P&L or customer outcomes in the next 12–18 months?
- Are we clear on *why* we are pursuing AI — growth, efficiency, risk reduction, or all of the above?
- Do our AI initiatives align with our broader business strategy and competitive positioning?

Data & Foundations

- What data debt blocks our top use cases, and how quickly can we fix it?
- Do we have a scalable data architecture and governance model to avoid silos?
- Have we embedded privacy, bias testing, & security into our data strategy from day one?

Execution & Ownership

- Do we have clear owners and sponsors for each prioritized use case?
- Have we set success metrics and kill/scale thresholds for pilots?
- Where must we be world-class (build) vs. where can we be good enough (buy/partner)?

Culture & Adoption

- Do employees understand how AI supports them rather than replaces them?
- Have we created training pathways and day-in-the-life playbooks for adoption?
- Are incentives aligned to reward usage of AI systems, not just old methods?

Scaling & Governance

- Are our AI projects designed as reusable capabilities, not one-off experiments?
- Do we have governance structures in place to scale responsibly?
- How will we monitor for drift, bias, and compliance once AI is deployed at scale?

Risk & Trust

- Do we have a plan for incident response if an AI system fails or causes harm?
- How will we ensure transparency and explainability in high-stakes decisions?
- Are we clear on what steps are we taking to build trust with customers, regulators, and employees?

APPENDIX 2: AI-POWERED DATA CAPTURE & UTILIZATION PLATFORMS

1. Snowflake

A cloud-native data warehousing and analytics platform that powers modern AI workloads. It excels at centralizing data from multiple sources, supporting AI model hosting, and scaling AI-driven analytics across enterprise environments. Enterprises view Snowflake as a cornerstone for AI strategy and data modernization.

2. Dataiku

A universal AI platform known for orchestrating the end-to-end life-cycle of analytics, models, and AI agents. It streamlines tasks from data preparation and model development to deployment and governance—empowering organizations to operationalize AI effectively.

3. Domo

An AI-embedded data and product platform that helps businesses connect, prepare, and explore data from any source. Domo accelerates decisions by offering AI assistance across the data journey—from ingestion to insight.

4. Sigma Computing

Offers an AI-first enterprise intelligence platform where teams can build interactive, generative AI-powered data apps directly on top of cloud data warehouses. It supports live data workflows, AI model invocation, and low-code analytics.

5. Qlik

A comprehensive data integration and analytics solution that brings together powerful AI features like natural language Q&A (Qlik Answers), AutoML for no-code predictive modeling, and multi-source data processing—all aligned around its associative analytics engine.

6. SnapLogic

An intelligent iPaaS (integration Platform as a Service) with AI-powered tools—such as Iris Integration Assistant and SnapGPT—to automate data pipeline creation, API integration, and AI agent orchestration within hybrid and multi-cloud environments.

7. Zapier

A no-code automation tool with AI-enhanced orchestration capabilities, enabling businesses to connect thousands of apps, automate workflows, and inject AI-driven agents to manage data-driven tasks and integrations at scale.

8. IBM watsonx

IBM's enterprise AI platform featuring components for data management (watsonx.data), model development (watsonx.ai), and AI governance (watsonx.governance). It enables businesses to train AI on proprietary datasets while ensuring compliance.

APPENDIX A: AI-POWERED FORECASTING AND BUDGETING PLATFORMS

1. Financial Planning & Analysis (FP&A) Focused Platforms:

- **Datarails**
  - An AI-augmented FP&A tool that automates data consolidation, budgeting, forecasting, scenario modeling, and visualization. Offers real-time financial insights and templates to streamline financial processes.
- **Workday Adaptive Planning**
  - Part of Workday's Business Planning cloud, this SaaS solution integrates AI for automated data consolidation, forecasting, budgeting, and collaborative planning. Seamlessly updates in real time.
- **Anaplan (PlanIQ)**
  - Provides AI-infused planning, budgeting, and forecasting capabilities within a broader Planning-Budgeting-Forecasting (PB&F) ecosystem. Built for enterprise financial planning transformation.

2. All-in-One FP&A Tools:

- **Cube**
  - Recognized among the top AI FP&A platforms of 2025. It helps with financial planning, budgeting, reporting, and forecasting using AI-based automation.
- **Vena, Pigment, Jirav, Fluence, SAP Analytics Cloud, Planful Predict**
  - Part of the top 10 AI FP&A tools for 2025, offering varied capabilities in forecasting, visualization, and planning.

3. Specialized Forecasting & Budgeting Tools:

- **IBM Planning Analytics (powered by TM1)**
  - A robust performance management suite enabling collaborative budgeting, forecasting, "what-if" analyses, and real-time data calculations via in-memory OLAP cubes —available on-premises or SaaS.
- **Jedox**
  - A Gartner-recognized leader in financial planning software, Jedox launched **JedoxAI**, introducing natural-language interaction to enhance business planning and analytics.
- **Martus, Prophix, Vena, Anaplan, IBM Planning Analytics:**
  - Highlighted as top-tier tools for budget forecasting across organizations—including nonprofits and SMEs.

4. Small Business and Spreadsheet-Friendly Options:

- **PlanGuru**
  - Budgeting and forecasting software popular with startups and SMEs; cost-effective, and highly rated in business-plan comparisons.
- **LiveFlow**
  - A spreadsheet-native solution (Excel & Google Sheets) that automates data sync, real-time forecasting, and dashboard updates.
- **Planguru** (not to be confused with PlanGuru)
  - Offers forecasting and budgeting tools tailored for small-to-medium businesses and nonprofits—starting at around $99/month.
- **Budgeto**
  - Designed specifically for small business financial planning and forecasting. Simplifies budgeting for users with limited finance expertise.

5. Cash Flow-Focused AI Forecasting:

- **DataRobot's Cash Flow Forecasting App**
    - Integrates with ERP systems like SAP to provide adaptive, real-time forecasting of cash flow. Helps finance teams anticipate working capital needs, reduce borrowing, and improve financial clarity. For example, King's Hawaiian cut interest expenses by over 20% using the platform.

APPENDIX B: AI-POWERED FRAUD DETECTION & RISK MANAGEMENT PLATFORMS

1. Feedzai

An AI-native platform offering real-time fraud and financial crime prevention across payment channels. It scores transactions instantly using a unified risk model to protect banking and payments environments.

2. Sift

Delivers AI-powered fraud decisioning at scale, protecting every stage of the digital user journey. One retailer experienced an 85% drop in chargebacks after deploying Sift.

3. Fraud.net (FraudNet)

An end-to-end enterprise fraud & risk intelligence suite that includes AI-based transaction monitoring, entity screening, case management, and anomaly detection. Companies report up to 97% fewer false positives and 80% reduction in fraud.

4. Sardine

Combines behavioral biometrics and device intelligence with data enrichment from multiple providers. It powers automated risk workflows, including KYC/KYB processes, with flexible rule-building and testing.

5. DataVisor

A SaaS platform for large organizations and financial institutions, enabling real-time AI-driven fraud and risk response. Powerful against sophisticated, evolving threats.

6. ComplyAdvantage

Specializes in detecting and managing AML and fraud risk using AI, machine learning, and NLP. Offers real-time transaction monitoring, payment screening, and sanctions detection.

7. Forter

A SaaS fraud prevention platform for e-commerce that fuses AI and behavioral analytics to authenticate and approve legitimate purchases while blocking fraudsters. Processes over $1 trillion in digital transactions.

8. ClearSale

Combines machine learning with expert human review to flag fraudulent transactions in e-commerce. Designed to reduce false positives while ensuring customer satisfaction.

9. Clari5 (CustomerXPs)

Now part of Perfios, this real-time financial crime risk management platform is used by banks for cross-channel fraud detection, AML compliance, and onboarding risk — plus its new generative AI tool, *Clari5 Genie.*

10. Quantexa

Offers a decision intelligence AI platform—featuring entity resolution, graph analytics, and network modeling—for fraud detection, risk analysis, and compliance across sectors.

11. HID Global's AI Risk Management

A behavioral intelligence-based solution designed to prevent digital threats like phishing, malware, account takeover, and more. Offers up to 90% fewer false positives and significant reductions in authentication costs.

12. Hawk.ai

An award-winning AML (Anti-Money Laundering) and CFT (Counter-Financing of Terrorism) platform using explainable AI. Combines transaction monitoring, payment screening, and fraud prevention to boost coverage and reduce false positives.

13. Oscilar

An AI Risk Decisioning™ platform that automates onboarding, fraud, credit, and compliance risk for consumer and merchant operations using adaptive, agentic AI models.

14. DataDome

Focused on bot detection and online fraud prevention, this ML-driven platform protects websites, apps, and APIs from automated threats like credential stuffing or DDoS attacks using device fingerprinting and behavioral analysis.

APPENDIX C: AI-POWERED EXPENSE OPTIMIZATION AND COST CONTROL

1. Expense Management

- **Expensify**

Offers SmartScan OCR, real-time fraud detection, AI-driven policy compliance, virtual "Concierge" assistant, and seamless integrations with major accounting systems. It significantly streamlines receipt capture, approvals, and reimbursement workflows, reducing manual effort and boosting accuracy.

- **Rydoo**

An AI-powered expense management solution trusted by over 1 million users. It enables mobile expense submission, automated approvals, fraud detection, and powerful insights—all integrated with ERP, HR, and finance systems.

- **Fyle**

Uses OCR and AI to extract data from receipts, auto-code expenses to the appropriate general ledger, and dramatically speed up receipt processing (up to 5× faster).

- **Yokoy**

Provides intelligent, automated expense management—simplifying expense submission and approval while ensuring compliance with company policies through AI-powered workflows.

2. AI-Native Spend & Procurement Management

- **Coupa**

A platform for total spend management, Coupa offers AI-driven insights across procurement, travel & expense, supply chain, and payments. Its Community.ai benchmarks company spending against anonymized industry data and supports generative AI tools like Coupa Navi to guide decisions and risk-informed recommendations.

3. Comprehensive Spend Automation & Intelligence

- **Airbase**

An all-in-one spend management system with AI-powered accounts payable automation. It auto-processes receipts and invoices, matches them to transactions, and categorizes them accurately—streamlining spend control and reducing manual workload.

- **Ramp**

A rapidly growing AI finance platform with corporate card and expense automation. Ramp's AI agents analyze transaction patterns, company policies, and even contextual data (e.g., from Gmail and Google Calendar) to categorize expenses, flag compliance issues, and automate approvals. Ramp is also developing a budgeting agent to proactively highlight potential issues to FP&A teams.

4. Cloud Cost Optimization Tools (FinOps-Oriented)

- **nOps**

An AI-powered FinOps platform designed to help organizations manage and optimize cloud usage, commitments, and overall spend—providing visibility and actionable insights for cloud cost control.

- **Cast.AI**

Leverages AI and automation to optimize cloud costs, manage resources efficiently, and support multi-cloud environments across AWS, Azure, and GCP—delivering real-time recommendations and DevOps collaboration tools.

- **Usage.AI, Pelanor**
  - **Usage.AI** uses machine learning for real-time cost tracking, budget controls, and intelligent instance resizing in cloud environments.
  - **Pelanor** offers dynamic cloud cost allocation and expense visibility—especially useful for Kubernetes-centric workloads.

5. Prescriptive & Supply Chain Cost Optimization

- **AIMMS**

A prescriptive analytics platform for advanced cost optimization across industries. It enables supply chain modeling, scenario planning, and "what-if" cost analysis—helping businesses reduce waste and inefficiencies.

APPENDIX D: AI-POWERED CASH FLOW MANAGEMENT PLATFORMS

1. HighRadius

A treasury-focused platform that uses **agentic AI**—autonomous AI agents that continuously monitor liquidity data, recalibrate forecasts, and flag anomalies. It enables adaptive, real-time cash flow forecasting with up to **95% accuracy** and cut manual forecasting effort by up to **70%**.

2. DataRobot Cash Flow Forecasting App

This tool integrates with SAP systems (e.g., S/4HANA Finance, SAP Datasphere) to deliver **adaptive, high-precision forecasting**, better anticipate cash availability, optimize working capital, and reduce reliance on short-term borrowing—used for example by King's Hawaiian to cut interest costs and improve forecasting clarity.

3. Kyriba (TAI)

Kyriba offers **AI-powered cash forecasting** within its broader treasury management suite. Kyriba's "TAI" (agentic AI) enhances real-time cash insights, liquidity planning, and forecasting—bridging ERP, payments, and banking data for improved accuracy and agility.

4. C3 AI Cash Management

Focused on banking and institutional use, this platform predicts client behaviors related to cash balances—such as identifying impending withdrawals up to 90 days in advance—and helps protect deposits and make targeted rate offers to retain liquidity.

5. Arya.ai Cash Flow Forecasting

Arya.ai's platform uses AI to enhance **forecast accuracy and simplify data reconciliation and tagging**, aiding scalable business forecasting and integration across finance systems.

6. OmniData's AI Recommendation Engine

OmniData employs predictive modeling to forecast customer and supplier payment behaviors, enabling proactive scenario planning like early-payment discounts and improving overall liquidity visibility and cash flow resilience.

7. Datarails – Cash Management for CFOs

Building on its FP&A and Excel-integrated planning platform, Datarails added a **Cash Management solution** to automate real-time consolidation, forecasting, and reporting—bringing AI-driven visibility directly into CFO workflows.

8. Affiniti – "AI CFO Agents" for SMBs

A fintech startup offering AI agents that act as digital CFOs for small and medium-sized businesses—overseeing cash flow, analytics, and banking tasks, even tailored to specific industries like healthcare and automotive.

9. Intuit AI Agents in QuickBooks

Intuit introduced AI agents integrated into QuickBooks to automate bookkeeping, payments, and financial workflows—providing **real-time insights** and improving small business cash flow management by reducing manual work by up to 12 hours per month.

10. JPMorgan's Cash Flow Intelligence

An internal, AI-driven tool for corporate clients that has reportedly **reduced manual work by nearly 90%** in cash flow analysis and forecasting. Currently used by around 2,500 clients, though not yet broadly commercialized.

APPENDIX E: AI-DRIVEN COMPREHENSIVE FINANCIAL MANAGEMENT PLATFORMS

1. Arya.ai — Broad, Modular AI Financial APIs

Arya.ai offers an extensive suite of AI APIs tailored for enterprise financial workflows. These include modules for **cash** flow forecasting, invoice processing, document fraud detection, risk compliance, and more—making it one of the more versatile options available.

2. DataRobot — Predictive Analytics Across Finance

DataRobot provides a robust predictive AI platform for financial use cases. It enables cash flow forecasting, budget modeling, anomaly detection (useful for fraud and expense irregularities), and automated analysis for cost trends.

3. NetSuite (Oracle) — Enterprise ERP with AI Enhancements

NetSuite offers a cloud-based ERP suite covering accounting, cash management, expense automation, plan & budget, and AI-driven document processing (via OCR) for invoices and receipts. Although it doesn't market itself explicitly for fraud detection, it offers anomaly flagging through AI and generative tools for financial narratives.

4. Planful — AI FP&A + Anomaly Detection

Planful's platform centers on forecasting, budgeting, and scenario planning, complemented by anomaly detection (which could flag potential fraud or unexpected cost behaviors). It doesn't directly address cash flow or expense optimization, though.

As of yet, there doesn't seem to be a single, off-the-shelf product that perfectly bundles all four capabilities with deep specialization in each. Undoubtedly, someone has deployed AI to assist in creating one, so stay on the lookout.

APPENDIX F: UPSKILLING & RETRAINING RESOURCES FOR FINANCE TEAMS EMBRACING AI

1. Executive Education & University Programs

- Wall Street Prep + Columbia Business School
  - *AI in Business & Finance Certificate Program (8 weeks, no coding required)* — Built for finance pros of all backgrounds, this program teaches AI proficiency step-by-step, culminating in a certificate from Columbia Business School Executive Education.
- University of Pennsylvania (Wharton)'s "AI for Business" Course
  - A flexible, online initiative (~4–6 weeks, ~2 hours/week) that covers AI fundamentals and how they apply across business and finance functions.

2. Specialized Online Courses & Certifications

- Corporate Finance Institute (CFI): AI for Finance Specialization
  - A hands-on program featuring nine modules—covering AI use in financial statement analysis, scenario planning, risk, dashboarding, and even prompt engineering.
- Udemy: AI Skills for Finance Professionals Learning Path
  - Self-paced training focused on prompt crafting, financial analysis, data insights, productivity, and risk management —all using AI.

3. Community & Cohort-Based Learning

- AI Finance Club by Nicolas Boucher
  - A peer-based learning community for finance professionals that includes live workshops, implementation-focused instruction, tool-specific frameworks, and monthly masterclasses. Tailored paths exist for beginners, intermediate, and advanced users.

4. Corporate & Internal Training Programs

- KPMG AI Training Sessions
    - KPMG teaches internal teams foundational AI skills—including prompting techniques like "chain of thought" and practical use cases such as automating audit memos with ~80% AI-assisted completion.
- Industry-Wide Shift in Learning Focus
    - A Financial Times panel and research team highlights the move towards AI literacy in finance programs at institutions like HEC Paris, Cambridge Judge, and Imperial College—melding machine learning, ethics, Python, and practical applications like trading strategies.
- PwC Australia's Upskilling Initiative
    - PwC now emphasizes "human skills" (critical thinking, ethics, collaboration) alongside AI training, delivering these via internal courses and platforms like Udemy, including AI strategy awareness.

Why These Matter

- Relevance to Finance Roles: Finance-targeted modules (e.g., scenario modeling, prompt engineering, AI agents for finance workflows) ensure high immediate applicability.
- Blended Formats: Options span self-paced courses, cohort learning, and live workshops—making upskilling flexible and scalable.
- Strategic Impact: CFOs and finance leaders are shown to benefit from both technical and soft skills as AI transforms high-value roles.
- Cultural Adoption: Companies like KPMG and PwC have shown that hands-on, guided learning accelerates adoption and encourages practical implementation.

APPENDIX G: AI-POWERED RECRUITMENT & SCREENING PLATFORMS

1. Candidate Sourcing and Matching

- **Juicebox (PeopleGPT)** – An AI search engine that scours over 800 million public profiles across 30+ sources to compile refined candidate lists tailored to specific qualifications.
- **Torre.ai** – A fully automated end-to-end recruiting platform using AI to locate, rank, and re-engage candidates—reportedly three times faster and 50% more cost-effective than traditional methods.
- **Findem** – A talent data platform that consolidates and surfaces top-of-funnel candidate activity from diverse datasets, enabling automated sourcing and outreach.

2. Applicant Tracking & Screening Automation

- **Zoho Recruit** – An SMB-friendly AI-enhanced ATS offering resume parsing, candidate matching, and customizable workflows, with seamless integrations.
- **Skillate** – Features an AI-driven matching engine that ranks applicants by relevance and helps automate and accelerate the recruitment process.
- **Manatal** – Scalable ATS sourcing candidates from 2,500+ channels, with automated pipelines and intuitive workflows.

3. Candidate Engagement & Interviewing

- **Paradox (Olivia)** – An AI chatbot that engages candidates via chat (SMS, web, mobile), pre-qualifies them, and schedules interviews. Useful for high-volume hiring environments.
- **HireVue** – A pioneer in AI-enabled video interviewing that evaluates verbal and facial cues (though it has received criticism for potential bias).

- **Metaview** – Automates interview note-taking, transcription, and candidate comparisons, helping recruiters focus on conversation rather than documentation.

4. Inclusive Job Description & Application Tools

- **Textio** – AI-enhanced writing assistant for crafting inclusive and effective job descriptions and communication—a tool that helps reduce bias in hiring language.

5. AI Assessment & Simulation Platforms

- **Canditech** – Offers cloud-based job simulations, cognitive and behavioral assessments, video interviews, and chatbot screening. Widely praised for fairness and candidate assessment accuracy.
- **Knockri** – Tools for skills-based assessments via video / audio responses, relying on transcript analysis to mitigate bias and improve transparency.

6. Internal Talent Matching & Mobility

- **Gloat** – An AI-driven internal talent marketplace matching employees to projects, mentorship, gigs, or roles based on competencies and career goals.

APPENDIX H: AI-POWERED EMPLOYEE ENGAGEMENT & RETENTION PLATFORMS

1. WorkStep

- **What it does**: An AI-driven platform for frontline workforce management. It analyzes open-text feedback (e.g., from surveys or chat), surfaces trending concerns, and offers actionable insights on a dashboard.
- **Impact**: Reported outcomes include a **20% reduction in turnover**, a **+5 boost in employee Net Promoter Score (eNPS)**, and a **3% improvement in safety sentiment**.

2. Culture Amp, 15Five, Officevibe

- **What they do**: These platforms deliver **pulse surveys**, performance check-ins, and engagement feedback powered by AI.
- **How AI helps**: They detect engagement trends, highlight drivers of low morale, and flag potential turnover risks in real time.
  - *Culture Amp* provides comprehensive dashboards, DEI analytics, and attrition risk prediction.
  - *15Five* combines pulse surveys with predictive models for engagement drivers.
  - *Officevibe* is known for simplicity, anonymity, and psychological safety—especially for smaller teams.

3. Humu (now part of Perceptyx)

- **What it does**: Sends personalized **behavioral "nudges"**—small, science-backed prompts aimed at improving employee happiness, performance, and retention.
- **How it helps**: Based on employee feedback and data trends, it encourages minute behavioral shifts that cumulatively enhance engagement.

4. Centrical

- **What it does**: A gamified employee engagement and performance platform combining microlearning, performance tracking, coaching, and AI-powered insights.
- **How it helps**: Uses gamification (points, badges, leaderboards) to motivate employees, while its AI dashboards help identify trends, performance gaps, and engagement levels.

5. Leena AI

- **What it does**: A conversational AI "virtual assistant" for employees—like Siri for internal use. It answers queries, surfaces HR policy info, and handles common administrative tasks.
- **How it helps**: By improving accessibility and clarity around workplace processes, it reduces frustration and fosters a smoother employee experience.

6. Workhuman's "Human Intelligence" Feature

- **What it does**: Augments its social recognition platform with AI to help craft meaningful praise messages and identify deserving employees for recognition or mentorship.
- **Impact**: Improves peer-to-peer recognition quality and helps managers spot high performers easily. Clients include BP, Cisco, and LinkedIn.

7. Lattice AI Agents

- **What it does**: Lattice is deploying proactive AI agents that handle HR queries, detect signs of disengagement, simulate difficult conversations, and support managers with real-time assistance.
- **Benefits**: Intended to reduce the burden of routine tasks and give leaders more bandwidth to focus on strategic workforce initiatives.

APPENDIX I: AI-POWERED WORKFORCE PRODUCTIVITY & TASK AUTOMATION PLATFORMS

1. Robotic Process Automation (RPA) & Process Automation

- **UiPath** – Leading RPA platform automating back-office tasks like data entry, CRM updates, and compliance.
- **Microsoft Power Automate** – Low-code tool for automating workflows across Microsoft 365 apps.
- **Automation Anywhere** – Enterprise-grade RPA platform for finance, HR, and customer operations.
- **Blue Prism** – Secure RPA for large-scale automation in regulated industries.
- **WorkFusion** – AI-driven automation platform for banking, finance, and compliance.

2. AI-Driven Employee Scheduling

- **Workday Workforce Scheduling** – Aligns worker preferences with business needs using AI-driven forecasting.
- **Rotageek** – Predicts labor demand and generates optimized shift schedules.
- **Workforce.com** – AI-powered scheduling with demand forecasting; reduces manual scheduling by 80%.
- **7shifts** – Restaurant-focused scheduling and payroll automation.

3. AI Workflow Assistants & Automation

- **Moveworks** – Conversational AI platform resolving IT/HR requests via Slack/Teams.
- **Deloitte MyAssist** – Enterprise AI assistant that automates internal workflows and document reviews.
- **Zapier AI** – Workflow orchestration platform connecting 8,000+ apps with AI chatbot and agent automations.

- **Monday AI** – AI features within Monday.com for task routing, content generation, and project workflows.

4. Productivity Analytics & Insights

- **ActivTrak** – Workforce analytics tool that identifies productivity trends and inefficiencies.

5. AI-Enhanced Productivity Suites

- **Motion** – All-in-one productivity suite with AI task management, scheduling, meeting notes, docs, and reporting.
- **Notion AI** – AI assistant embedded in Notion workspace for drafting, summarizing, knowledge search.
- **Asana AI Studio** – Customizable AI agents embedded in Asana to automate project workflows.

APPENDIX J: AI-POWERED SKILLS-GAP ANALYSIS PLATFORMS

1. iMocha

An AI-powered skills assessment and intelligence platform that aggregates skills data to highlight strengths and gaps across departments. It supports reskilling, succession planning, and talent benchmarking. Pricing starts at around $1,800/year for basic plans, with higher tiers offering more advanced features

2. TechWolf

Provides **real-time** visibility into skill supply versus demand across your workforce. It integrates seamlessly with existing HR systems (HRIS, ATS, LMS, etc.) to surface critical gaps and guide hiring, internal mobility, or reskilling decisions. The platform includes dynamically updated skill taxonomies and actionable dashboards.

3. Spire.AI

Offers AI-driven "SkillAI™" that performs adjacency analysis—finding logical skill transition paths for employees—and recommending personalized learning pathways. It also helps with internal redeployment and talent matching.

4. JobsPikr

Uses real-time labor market and workforce data to pinpoint critical skill gaps. It's especially useful for aligning training and recruitment with emerging business needs—serving over 1,000 companies globally.

5. SkyHive

A cloud-based reskilling and workforce management tool that matches employees' current skills to both current and future job requirements. It also maps career pathways and suggests training to internal mobility.

6. Disprz

An AI-driven learning and skilling platform (LMS + LXP + skill analytics). It uses generative AI and custom assessments to power skill analytics and personalized learning for millions of users.

7. Workera's Sage

An AI agent specialized in deeply contextual skill assessments with tailored learning plans designed to align individual and organizational objectives.

8. Aura Intelligence

Transforms raw HR data into actionable insights by identifying internal skill gaps through analytics of performance reviews, sentiment data, knowledge assessments, and more.

9. IBM Watsonx & TalentGuard

Highlighted by Robert Half as practical AI tools for HR that help identify skill gaps by analyzing employee data and comparing it with future needs.

APPENDIX K: AI-POWERED CUSTOMER DATA PLATFORMS (CDPS)

1. Customer Data Platforms (CDPs) specialize in collecting, unifying, and activating customer data across channels, enhanced by AI for better personalization and insights.

- **Adobe Experience Platform**: A leading AI-driven CDP widely used in enterprise-grade marketing stacks.
- **Amperity**: Leverages AI to perform robust identity resolution, building accurate Customer 360 profiles and automating data workflows—with support for data security standards like SOC 2, GDPR, and HIPAA.
- **Bloomreach, Dotdigital, Informatica, Insider, Lytics, Ortto, Segment, Terminus Data Studio**: Other strong AI-enabled CDP options listed in the Top 10 for 2025.
- **Treasure Data**: A sophisticated AI-powered "Intelligent CDP" offering features like the "Diamond Record" (combining first-, second-, and third-party data) and real-time AI decisioning, widely adopted by Global 2000 brands.
- **Tealium**: Provides real-time customer insights via a trusted, scalable CDP compatible with a wide range of enterprise tech stacks.

2. AI-Enhanced CRMs & Customer Relationship Tools

CRM platforms infused with AI help with automation, forecasting, personalization, and operational efficiency.

- **Salesforce (Einstein & Agentforce)**: Salesforce's AI suite delivers predictive and generative AI across the Customer 360 —including sales forecasting, email automation, conversational agents, and more.
- **Zoho CRM (Zia)**: Features AI assistant "Zia"—adds real-time insights, anomaly detection, workflow suggestions, and AI agents for support, sales, and marketing via the Zia Agent Marketplace.

- **SuperAGI's AI-CRM Platform**: Uses AI agents to automate inbound/outbound sales, journey orchestration, and omnichannel marketing workflows.

3. Additional AI Tools with Customer Data Focus

- **SnapLogic**: A data integration and automation platform with agentic AI capabilities, SnapGPT generative copilot, and AI-powered workflow orchestrators—helpful for cleaning and funneling customer data.
- **Synerise (AI Growth Cloud)**: Provides unified customer data analytics, predictive behavior modeling, scoring, segmentation, and automated marketing (email, SMS, in-store) based on both online and offline data.

Why AI matters in CDPs: AI enhances identity resolution, cleans and merges duplicative records, enables predictive segmentation, personalized journey orchestration, and next-best-action recommendations.

APPENDIX L: AI-POWERED PERSONALIZED MARKETING PLATFORMS

1. Leading CDPs & Personalization Suites

- **Bloomreach**
  - Powered by its Loomi AI, Bloomreach enables real-time, highly tailored personalization—shaping customer experiences with dynamic product recommendations, content, and automated journey orchestration.
- **Adobe Experience Cloud**
  - A comprehensive suite for personalization at scale—delivering AI-driven recommendations, individualized site content, and optimized customer journeys via tools like Adobe Target, Campaign, and Audience Manager.

2. AI-Driven Ad & Campaign Optimization

- **Smartly.io**
  - Automates cross-platform advertising with AI—providing scaled campaign execution, creative automation, and intelligent performance insights.
- **StackAdapt**
  - An AI-native marketing platform that optimizes ad targeting, personalizes user experiences, and orchestrates campaigns across advertising and email channels with real-time insights.
- **Omneky**
  - Uses deep learning to generate creative ads, test performance variables, and launch omnichannel campaigns—automating personalized ad creative generation and deployment.

3. AI Marketing Agents & Campaign Creation

- **Auxia**
  - An emerging AI marketing startup using AI agents to personalize consumer shopping journeys—dynamically tailoring marketing content across customer touchpoints.
- **Adobe's AI Agents**
  - Adobe has introduced AI "agents" (e.g., via the Agent Orchestrator) that automate content generation, journey optimization, and campaign workflows—already leveraged by brands like Coca-Cola and Marriott.

4. AI Copy & Creative Generation Tools

- **Anyword**
  - An AI language platform that generates and optimizes marketing messaging—tailored for websites, social media, email campaigns, and ads—used by brands like National Geographic, Red Bull, and BBC.

APPENDIX M: AI-POWERED CUSTOMER SERVICE PLATFORMS

1. Zendesk

A widely adopted cloud-based customer support platform, Zendesk now includes powerful AI enhancements. It supports omnichannel support—chat, email, social, voice—and uses AI to triage tickets intelligently and reduce average resolution time by 30–60 seconds per ticket.

2. Freshdesk

An AI-powered helpdesk solution providing omnichannel support, predictive analytics, and automated workflows to streamline service operations and boost response efficiency.

3. LiveChat

A long-established conversational helpdesk platform combining AI chat, ticketing, and analytics to offer a unified customer service experience across web and mobile.

4. Intercom

A modern, all-in-one conversational support platform integrating AI chatbots, help desk functionality, and proactive messaging to enhance customer engagement and case resolutions.

5. Tidio

An AI-enhanced live chat and chatbot tool that integrates GPT-4 capabilities—perfect for website support and conversational automation.

6. Help Scout

AI-driven support software offering inbox management, automated workflows, summaries, agent assistance, and sentiment analysis—designed to improve agent productivity for small-to-mid-size teams.

7. Ada

A conversational AI platform focused on automation and personalization. It uses intelligent chatbots to deliver fast, multilingual customer support and automate key workflows.

8. Kommunicate

A generative AI-driven customer service automation tool that supports chat, email, and voice. Features include multi-channel integration, AI agent builders, language translation, summaries, and AI agent suggestions for support reps.

9. Sprinklr AI+

An omnichannel customer service platform harnessing generative AI (via Google's Vertex AI and OpenAI's models) to automate responses and provide team-level performance insights, all backed by strong data governance.

10. Crescendo.ai

An enterprise-grade AI support automation platform combining voice assistants, live chat, sentiment tracking, email ticketing, CSAT scoring, and knowledge base management—all with high accuracy and compliance features.

11. Conversica

Offers AI-powered virtual assistants that engage customers through two-way conversations via chat, email, and SMS—commonly used for sales, onboarding, and customer support automation.

12. Amelia (formerly IPsoft)

A sophisticated conversational AI agent that automates customer support and assists contact center agents during live interactions using deep natural language understanding.

13. Yellow.ai

A global automation platform offering AI-driven customer support across over 135 languages and 35+ channels—covering both chat and voice use cases.

APPENDIX N: AI-POWERED PLATFORMS WHICH OFFER BOTH CAMPAIGN MANAGEMENT AND PREDICTIVE ANALYSIS

1. Salesforce Marketing Cloud (Einstein AI)

- **What it offers**: A premier marketing automation suite integrated with AI capabilities through **Einstein**, which delivers predictive forecasting (e.g., customer churn risk, email engagement), personalized content delivery, and cross-channel campaign orchestration.
- **Best for**: large enterprises seeking advanced automation, data-driven decision-making, and seamless integration with CRM data.

2. Optimove

- **What it offers**: A relationship marketing platform powered by a built-in Customer Data Platform (CDP) delivering predictive customer modeling, micro-segmentation, real-time triggers, and multi-channel campaign automation—all managed via an intuitive calendar interface.
- **Best for**: Brands focused on personalized retention marketing across email, push, SMS, web, and social ads.

3. MoEngage

- **What it offers**: A customer engagement platform featuring AI-driven personalization, user analytics, real-time journey orchestration, and automation across multiple touchpoints like email, SMS, push, WhatsApp, and in-app messages
- **Best for**: Consumer-facing brands needing cross-channel automation with adaptive messaging and predictive insights.

4. Improvado AI Agent

- **What it offers**: Embedded in Improvado's analytics platform, this AI agent provides instant trend analysis, real-time visualizations, and benchmarking via natural language queries —accelerating data-driven campaign optimization.
- **Best for**: Marketing teams who want conversational, predictive insights from their campaign data without deep analytics expertise.

5. Adverity

- **What it offers**: A unified marketing data analytics platform that centralizes campaign data for visualization, reporting, and analytics. While not explicitly labeled "predictive," it supports sophisticated insights and trend analysis.
- **Best for**: Data-driven teams managing diverse campaign sources needing streamlined analytics workflows.

6. Supermetrics (Marketing Intelligence Platform)

- **What it offers**: Automates data collection from various media and analytics sources into BI tools (e.g., Looker Studio, Power BI), enriched by AI-powered data transformation and real-time insights for activation.
- **Best for**: Marketing teams requiring unified performance reporting and intelligence across channels.

7. Microsoft Azure ML & Google Cloud AI Platform

- **What they offer**: Robust AutoML platforms enabling marketers and analysts to build and deploy predictive models using campaign data—ideal for forecasting behavior, performing lead scoring, and optimizing segmentation.
- **Best for**: Teams with data science capabilities or those already embedded in Microsoft or Google ecosystems.

APPENDIX O: AI-POWERED PRICING & SALES OPTIMIZATION PLATFORMS

AI-Powered Pricing Optimization Tools

1. Pricefx

A leading enterprise-grade solution for pricing science—offering modules for price optimization, price management, promotions, rebates, and CPQ (Configure-Price-Quote). It consistently ranks highly on G2 for ease of use and ROI.

2. PROS

Specializes in AI-driven pricing, revenue management, and CPQ tools. Ideal for industries like travel, manufacturing, and enterprise B2B where dynamic pricing and margin control are critical.

3. Vistaar

A SaaS provider focused on price, promotion, configuration, and rebate optimization. It uses predictive models and machine learning to help businesses grow profitably across diverse industries.

4. Intelligence Node, Omnia Retail, Competera

These tools serve retail and e-commerce with AI-powered **dynamic pricing** capabilities—adjusting prices in real time based on demand, competitors, inventory levels, and sales performance. They help retailers increase revenue (by ~10%) and reduce costs (~5%).

5. Revionics

A retail-focused pricing analytics platform offering AI-driven base pricing, markdown strategies, promotions, and lifecycle pricing optimization—lauded by major retailers for precision and performance.

6. Pricen (by Sniffie)

Offers reinforcement learning-based pricing optimization. It supports clever product grouping to prevent price cannibalization and claims to reduce manual pricing effort by up to 95%.

AI-Powered Sales Optimization Tools

7. Salesforce Einstein (Sales Cloud)

Einstein adds AI-driven predictive insights to Salesforce's Sales Cloud —helping forecast revenue, prioritize leads, and recommend next best actions. Highly integrated for CRM and sales operations.

8. HubSpot Sales Hub

A CRM and sales engagement tool enhanced with AI—supports lead qualification, deal forecasting, and automated outreach across channels, making it accessible and scalable for many businesses.

9. Optimove

While often known for marketing, Optimove is worth mentioning because of its emphasis on predictive customer modeling and personalized campaign orchestration. Effective in boosting sales through retention marketing.

10. Docket

An AI-powered revenue platform acting as a virtual sales engineer and "AI seller." It provides real-time technical support during deals, automates documentation (like RFPs), and engages website visitors via conversational AI—boosting win rates by ~12% and slashing operational overhead by ~83%.

11. Conversica AI Assistants

Intelligent virtual assistants that handle customer outreach—via email, chat, and SMS—to qualify leads, schedule meetings, and automate parts of the sales funnel.

Industries with Specialized AI Pricing Tools

1. Retail & E-Commerce

- **Examples:** Revionics, Competera, Omnia Retail, Intelligence Node, Sniffie.
- **Focus:** Dynamic pricing, markdown optimization, promotion planning.

- **Why specialized:** Retailers deal with high SKU counts, frequent promotions, seasonality, and competitor-driven price sensitivity. AI here tracks competitors, demand patterns, and inventory levels in real time.

2. Travel & Hospitality

- **Examples:** PROS, RateGain, Amadeus AI.
- **Focus:** Revenue/yield management, fare optimization, personalized offers.
- **Why specialized:** Airlines and hotels rely on capacity-based, perishable inventory. AI adjusts prices dynamically based on booking windows, demand forecasts, and competitor pricing.

3. Manufacturing & B2B

- **Examples:** Pricefx, Vistaar, Zilliant, PROS (B2B module).
- **Focus:** Configure-Price-Quote (CPQ), contract pricing, channel partner pricing, margin optimization.
- **Why specialized:** B2B pricing is highly contract-driven, negotiated, and complex, with multiple tiers and discount structures. AI here balances margin protection with deal win rates.

4. Energy & Utilities

- **Examples:** Zilliant, Vendavo.
- **Focus:** Tariff optimization, contract pricing, demand-based adjustments.
- **Why specialized:** Utility and energy pricing must consider regulations, demand forecasting, and regional consumption patterns.

5. Consumer Packaged Goods (CPG) & FMCG

- **Examples:** NielsenIQ Revenue Growth Management, Revionics.

- **Focus:** Trade promotion optimization, price pack architecture, retailer collaboration.
- **Why specialized:** CPG pricing depends heavily on promotions, retailer relationships, and optimizing shelf pricing for fast-moving goods.

6. Pharma & Healthcare

- **Examples:** Model N, IQVIA Pricing Solutions.
- **Focus:** Market access pricing, rebate management, compliance.
- **Why specialized:** Pricing is highly regulated, with rebates, insurance reimbursement, and compliance considerations.

APPENDIX P: AI-DRIVEN PRICING & SALES PERSONALIZATION PLATFORMS

1. Dynamic Yield

A personalization and decisioning engine enabling AI-powered content and product recommendations across web, mobile, email, and ads. It leverages user affinities, intent, and real-time behavior to optimize pricing suggestions and customer journeys dynamically.

2. Bloomreach

A digital experience platform specializing in e-commerce personalization, product discovery, and content management. Its AI-driven "Discovery" component powers personalized recommendations, search experiences, and pricing triggers across digital channels.

3. Omneky

An AI-first platform for automating ad creative generation and omnichannel campaign execution. It analyzes performance data to personalize ad content—potentially including price-related messaging —and launches campaigns across channels like Meta, Google, and TikTok.

4. Autobound.ai

An AI outbound sales platform that scales personalized outreach by crafting insight-driven, brand-safe email communication. While pricing personalization per se isn't the primary function, its hyperpersonalized messaging can be adapted to incorporate price propositions or tailored sales offers.

5. AI Personalization via CRMs

Leading CRM systems such as **Salesforce, HubSpot, Zoho**, and **Freshsales** embed AI to enhance pricing and sales personalization—through predictive lead scoring, dynamic segmentation, behavior-driven messaging, and adaptive content.

APPENDIX Q: AI-POWERED SALES FORECASTING PLATFORMS

1. Aviso

A sales-focused forecasting solution combining AI with human input for heightened accuracy. It boasts forecasts with up to 98% accuracy, supporting unified predictive modeling.

2. Revcast

A next-gen "4-dimensional" platform that goes beyond deal-based forecasts, integrating data on team capacity, pipeline progression, and rep performance. It features AI co-piloting for scenario modeling, proactive alerts, and high-level revenue planning alignment.

3. Clari

Widely recognized for its revenue operations and forecasting intelligence. Clari provides predictive insights into pipeline health, deal progression, and forecast accuracy—making it a staple for GTM teams.

4. InsightSquared

A no-code analytics tool with deep AI forecasting capabilities—offering over 350 prebuilt dashboards and reports to monitor pipeline value and projected bookings.

5. Forecastio

Designed especially for HubSpot users, this tool integrates seamlessly and uses historical data to automate sales stage probabilities and provide multi-method forecasting insights. Great for SMBs relying on HubSpot.

6. Revenue Grid

A revenue intelligence platform capturing data from emails, chats, calendars, and calls to offer real-time pipeline tracking and AI-generated forecasting insights. Ideal for organizations seeking end-to-end visibility.

7. Salesforce Einstein (Sales Cloud)

The AI layer embedded in Salesforce's CRM suite. Einstein delivers predictive lead scoring, opportunity likelihoods, revenue forecasting, and tailored recommendations directly within your sales workflows.

8. HubSpot Sales Hub

An all-in-one CRM with built-in AI features for deal tracking, forecasting, and personalized outreach—well-suited for SMBs and growing sales teams.

9. Spotio

A field sales tool featuring an AI Sales Assistant that provides predictive pipeline insights, helping reps understand which deals are most likely to close and where to focus their efforts.

APPENDIX R: AI-POWERED REVENUE MANAGEMENT & PROFIT OPTIMIZATION TOOLS

1. PROS Smart Price Optimization & Management

An enterprise-grade tool that delivers **AI-powered, real-time pricing**. It helps businesses centralize pricing data, protect margins, and dynamically optimize prices across digital and negotiated sales channels. Users have reported up to **20% revenue uplift** and **5% margin improvement**.

2. IMA360

Specializes in **AI-driven profit optimization**, particularly for B2B industries. It automates pricing, rebates, promotions, and ship-and-debit strategies—helping businesses scale profitably with industry-specific configurations.

3. Duetto (Revenue & Profit Operating System)

A hospitality-centric platform consolidating pricing, forecasting, reporting, group optimization, loyalty, and profit insights under one system, enabling streamlined revenue management.

4. RealPage AI Revenue Management

Tailored for the real estate and property management sector, it empowers rental revenue optimization through data-driven AI models —reducing vacancy, improving performance, and aligning pricing strategy with asset goals to consistently outperform the market by 2–4%.

5. Atomize, IDeaS, LodgIQ, FLYR

Targeting hospitality and lodging industries, these platforms leverage machine learning to automate dynamic pricing and inventory optimization:

- **Atomize**–ML-driven pricing recommendations for rapid adaptation to market changes

- **IDeaS Revenue Solutions**–Advanced analytics and forecasting for better pricing decisions
- **LodgIQ**–AI-powered pricing and inventory optimization
- **FLYR Hospitality**–Strategic revenue optimization focused on rate and demand modeling

6. Revenue.ai

A platform offering AI-powered copilot and dashboard studio capabilities for pricing and revenue decisions. It aims to democratize access to insights and accelerate execution across teams.

7. o9 Solutions – Revenue Growth Management

This AI-enabled platform supports smarter, faster, profitable decisions with:

- Enhanced demand forecasts
- Price elasticity modeling
- Promotion predictions
- Real-time optimization and simulation

APPENDIX S: AI-POWERED DEMAND FORECASTING & INVENTORY OPTIMIZATION PLATFORMS

1. Blue Yonder (formerly JDA)

A global leader in supply chain planning, Blue Yonder offers integrated AI-driven demand and supply planning, omnichannel inventory optimization, and logistics management—all designed to elevate forecast accuracy and streamline operations.

2. Oracle SCM Cloud

Oracle's AI-powered SCM system delivers predictive analytics for demand forecasting, real-time inventory visibility, and replenishment optimization. Notably reduces inventory costs and improves supply chain efficiency by **10–15%**.

3. Logility / Slimstock / Relex / Netstock / GMDH Streamline / EazyStock

These tools are highlighted by industry reviews as top inventory forecasting platforms for e-commerce and retail, offering features like multi-channel data syncing, automated reorder recommendations, and performance dashboards.

4. Prediko

Designed for Shopify and e-commerce brands, Prediko provides SKU-level AI demand forecasting, real-time stock alerts, one-click purchase orders, and PO management across multiple stores and warehouses—starting at about **$49/month**.

5. Monocle AI

An AI-powered tool offering SKU-level demand plans, reorder suggestions, automated pattern detection, and forecast confidence scores—starting around **$47/month**

6. Inventory Forecasting Hero

A straightforward, budget-friendly solution (starting at **$25/month**) for Shopify users, giving customizable reorder forecasts, low-stock alerts, lead-time management, and inventory depletion timelines.

7. Infor Demand Forecasting

Features ML-driven demand sensing, exception-based alerts, near real-time insights, and rapid deployment—in as little as 30 minutes when paired with Infor ERP systems.

8. Intuendi (with AI assistant Symphonie)

This platform provides omni-channel demand planning, inventory replenishment, SKU/warehouse balancing, forecast generation, and AI support via its assistant Symphonie to orchestrate stock levels and orders.

9. StockIQ

A supply chain planning suite offering user-friendly AI-powered demand forecasting, simplified ordering, and inventory control through advanced machine learning models.

APPENDIX T: AI-POWERED LOGISTICS & TRANSPORTATION OPTIMIZATION PLATFORMS

1. Blue Yonder

Formerly JDA Software, Blue Yonder is a global leader offering a comprehensive AI-powered supply chain platform. It includes demand & supply planning, warehouse and transportation management, and logistics optimization, all enhanced by machine learning.

2. Kinaxis Maestro (formerly RapidResponse)

Kinaxis Maestro enables real-time supply chain planning and orchestration with AI-driven demand forecasting, scenario modeling, and risk management capabilities.

3. Optimal Dynamics

Specializing in AI-driven fleet management, Optimal Dynamics improves operational efficiency by matching freight loads, trucks, and routes. It delivers up to 24% increased weekly truck revenue for freight operators using its machine learning models.

4. Transmetrics

Transmetrics offers AI analytics and planning for logistics and trucking firms. Its solution covers linehaul planning, container management, fleet optimization, and enhanced profit margins.

5. V2T Logistics AI

An AI-driven Transport Management System (TMS) specifically for route planning, V2T Logistics AI optimizes delivery and collection routes—including last-mile logistics, depot coordination, and multi-stop planning—for reduced fuel cost and improved efficiency.

6. Shipsy

Shipsy specializes in AI-fueled logistics management, offering real-time visibility, predictive analytics, and automation. It easily integrates with TMS and ERP systems to streamline operations.

7. ClickPost

Focused on e-commerce logistics intelligence, ClickPost provides shipment tracking, returns management, carrier optimization, and logistics analytics. Recently, it added an AI voice agent, "Parth", to better manage failed deliveries.

8. Onfleet

Onfleet is a last-mile delivery platform that manages driver coordination, route optimization, and real-time tracking—frequently called the "Uber for delivery" and used by retail and food delivery businesses globally.

9. Optym

Optym offers optimization, simulation, and analytics solutions across transportation sectors—rail, trucking (especially LTL), air freight, and mining. Notable products include *RailMAX* for train scheduling and *HaulPLAN* for smart trucking operations.

APPENDIX U: AI-POWERED RESOURCE ALLOCATION & WORKFORCE SCHEDULING PLATFORMS

1. Mosaic

An AI-driven resource planning and management platform that automatically aligns staff with projects in real-time. It offers workforce capacity planning, workload balancing, project budgeting, and forecasting. Notably praised for reducing burnout and enhancing profitability through utilization optimization.

2. Wrike (Wrike Copilot)

A leading enterprise work-management tool enhanced with AI through **Wrike Copilot**, enabling teams to analyze and act on real-time work data. It offers insights for better resource distribution, task assignment, and capacity planning.

3. Dayshape

An AI-powered resource management tool that assists in scheduling, task allocation, and resource engagement—especially useful for professional services and accounting firms handling complex client projects.

4. Motion

Described as an AI project manager, Motion automates resource allocation and capacity planning—particularly effective in keeping projects on track and freeing teams from manual coordination, as reported by users like MP Cloud and Ally Advantage.

5. Workday Scheduling

Part of Workday's workforce management suite, this AI-powered module generates smart schedules by aligning worker preferences with business demand. Includes features like demand forecasting, labor optimization, and mobile-friendly preference settings.

6. IFS Workforce Scheduling & Planning

IFS offers real-time, AI-optimized workforce planning and scheduling software, ideal for field service environments. It ensures the right technician—with the right skills and parts—is dispatched to resolve service tickets efficiently and meet SLAs.

7. Shyft (Smart Resource Allocation)

Shyft's AI resource allocation system helps businesses optimize employee scheduling by analyzing demand patterns, improving staffing accuracy, cutting labor costs, reducing overtime, and enhancing employee satisfaction and customer service quality.

8. Workday Illuminate (AI Agents)

Workday has developed AI agents (via Illuminate and its Agent System of Record) that can assist across HR and workforce planning—including tasks like scheduling and resource allocation—ensuring operational alignment and governance across the digital workforce.

APPENDIX V: AI-POWERED SUPPLIER RISK MANAGEMENT PLATFORMS

1. Resilinc

An advanced **Agentic AI-driven** platform for supplier risk management featuring tools like Multi-Tier Mapping, EventWatch AI, and RiskShield. It delivers real-time supply chain visibility, predictive risk alerts, and tailored mitigation recommendations. Resilinc was named a Leader in the 2025 Gartner Magic Quadrant for supplier risk management solutions.

2. Interos

Claims to be the first fully automated supplier resilience platform, harnessing AI to map and monitor supply chains at scale—identifying risks earlier, deeper, and more comprehensively than traditional methods.

3. Everstream Analytics

Delivers AI-driven risk intelligence to help businesses build more agile, risk-proof supply chains. It emphasizes real-world intelligence to anticipate disruptions and optimize supply chain resilience.

4. Prewave

Utilizes AI and predictive analytics to transform millions of risk events into clear, actionable alerts—boosting supply chain cost-effectiveness and decision agility.

5. Apex Analytix

Offers AI-enabled supplier risk solutions with continuous monitoring, automated risk scoring, and insights aimed at detecting and resolving issues before they escalate.

6. Certa

Provides scalable AI automation with AI agents that gather data, assess risk, and generate compliance reports across thousands of third parties. Its system adapts workflows to policy changes using natural language configurations.

7. Vanta (Vendor Risk Management)

Enhances vendor security reviews with AI-powered automation— aids continuous monitoring, risk scoring, and compliance workflows, reducing review times by up to 50%.

8. UpGuard

Streamlines vendor risk assessment using AI—automating evidence collection, assessing inherent risks, and boosting risk workflow scalability and speed.

9. Panorays

A SaaS solution focused on third-party security risk. It automates vendor questionnaires, external attack surface reviews, and compliance assessments (e.g., GDPR, CCPA), supporting continuous monitoring.

10. Xapien

An AI-based due diligence platform built for regulated sectors, offering comprehensive third-party assessments using NLP and machine learning. It accelerates customer and vendor onboarding while bolstering compliance, including AML and reputational checks.

11. RepRisk

Offers ESG-focused supplier risk monitoring using AI and machine learning to analyze over 100,000 daily sources—tracking environmental, social, and reputational risks across a vast set of suppliers and projects.

APPENDIX W: SOME OF THE MOST COMMONLY USED AI TOOLS IN MANUFACTURING AND LOGISTICS

1. Predictive Maintenance & Equipment Monitoring

- **Uptake** – AI-based predictive maintenance and asset reliability.
- **SparkCognition** – machine learning for industrial predictive analytics.
- **C3 AI Reliability** – predictive maintenance for heavy equipment.
- **IBM Maximo** – asset performance monitoring with AI insights.

2. Supply Chain Optimization & Planning

- **Blue Yonder (formerly JDA)** – AI-driven supply chain planning, demand forecasting, and logistics optimization.
- **o9 Solutions** – end-to-end supply chain AI platform.
- **Llamasoft (Coupa AI)** – predictive modeling and scenario planning.
- **Kinaxis RapidResponse** – demand and supply balancing with AI.

3. Robotics & Automation

- **ABB Robotics AI, FANUC, Universal Robots (UR+)** – smart robotics for assembly, welding, and packaging.
- **GreyOrange** – AI-driven warehouse automation robots.
- **Locus Robotics** – autonomous mobile robots (AMRs) for logistics.

4. Quality Control & Computer Vision

- **Landing AI** – computer vision for defect detection.
- **Instrumental** – AI-driven product inspection in manufacturing.

- **Qualitas Technologies** – visual quality assurance.

5. Logistics & Transportation Optimization

- **ClearMetal (Project44)** – AI-powered demand and logistics visibility.
- **FourKites** – real-time logistics tracking and predictive ETAs.
- **Shipwell** – AI-driven freight management and route optimization.
- **Optibus** – AI optimization for fleet and transportation scheduling.

6. Inventory & Warehouse Management

- **Vecna Robotics** – AI-enabled autonomous material handling.
- **Berkshire Grey** – robotics for warehouse picking, sorting, fulfillment.
- **6 River Systems (Shopify)** – AI-powered warehouse automation.

7. Decision Support & Industrial AI Platforms

- **Siemens MindSphere** – IoT + AI analytics for industrial operations.
- **GE Predix** – industrial IoT and AI for factory operations.
- **Microsoft Azure IoT / AI** – predictive analytics for manufacturing.
- **Google Cloud Manufacturing AI** – demand prediction, defect detection, and process optimization.

APPENDIX X: SOME OF THE MOST COMMONLY USED AI TOOLS IN HEALTHCARE

1. Clinical & Diagnostic AI Tools

- **Aidoc**: A widely adopted AI radiology platform that automatically analyzes medical imaging like CT scans to flag urgent cases—such as strokes or pulmonary embolisms—for prompt review. With multiple FDA and CE approvals, it's integrated in over 900 hospitals globally, offering high sensitivity and specificity in detecting critical conditions.
- **PathAI**: This deep learning-powered tool enhances pathology slide interpretation, improving accuracy in diagnosing diseases like cancer and reducing human error.
- **Butterfly iQ & Caption Health**: Portable, AI-enabled ultrasound devices. Butterfly iQ attaches to a smartphone to guide users in capturing and interpreting images, ideal for remote or point-of-care settings. Caption Health offers real-time probe guidance for clinicians with limited imaging experience.
- **DeepMind Health**: Developed by Google's DeepMind in partnership with Moorfields Eye Hospital, this AI system analyzes retinal scans to detect conditions such as diabetic retinopathy and age-related macular degeneration with accuracy comparable to expert clinicians.

2. Clinical Documentation & Workflow Automation

- **DAX Copilot (Dragon Ambient eXperience)**: Developed by Nuance, this tool uses ambient AI to transcribe patient visits and auto-populate electronic health records (EHRs), greatly reducing documentation time. It integrates with systems like Epic.

- **Heidi Health**: An Australian AI medical scribe solution that transcribes clinical encounters into structured notes and summaries. It integrates with major EHR platforms and helps alleviate administrative burdens on providers.
- **Sully.ai**: Offers a suite of AI "medical employees" including AI Receptionist (patient intake), AI Scribe, AI Interpreter (multilingual support), and AI Medical Assistant (draft assessments). These tools automate routine tasks and integrate with EHR systems.

3. Clinical Decision Support & Information Retrieval

- **Merative (formerly IBM Watson Health)**: An AI platform for analyzing large volumes of clinical data—such as patient records, guidelines, and research—to support diagnostic and treatment decisions. Watson notably supports oncology and analytics-driven decision-making.
- **OpenEvidence**: A physician-only search engine delivering fully cited medical answers in seconds, used daily by ~40% of U.S. clinicians and leveraged in over 10,000 hospitals. Its AI agent—DeepConsult—summarizes hundreds of studies for complex medical questions.

4. Generative AI & Chatbot Applications

- **Ada Health**: A self-service diagnostic chatbot that asks users health-related questions and offers personalized assessments or guidance to appropriate care resources.
- **ChatGPT, Claude, Doximity GPT**:
  - *ChatGPT* and *Claude* are large language model (LLM) chatbots used for note summarization and drafting clinical content—though they lack built-in privacy protections.
  - *Doximity GPT* offers similar functionality but with HIPAA-compliance measures, making it more viable for clinical documentation.

- **Aiddison** and **BioMorph**: AI tools for **drug discovery**. Aiddison, from Merck, uses ligand- and structure-based methods to identify candidate molecules; BioMorph applies predictive analytics on compound datasets to accelerate drug development.

5. Emerging & Specialized AI

- **Ambient & Voice AI**: Tools like **DAX**, **Lyrebird Health**, and voice agents like **Eva** and **Infinitus** help with tasks such as automated documentation, insurance communications, and providing companionship to older adults.
- **AI in Acute Care**:
  - AI tools like an advanced **AI stethoscope** detect heart conditions (e.g., atrial fibrillation, valve disease) in ~15 seconds with high accuracy, though integration into practice still faces challenges.
  - AI-driven diagnostic software for stroke analysis has significantly sped up care, tripling full recovery rates by enabling faster interventions.
  - **DrugGPT**: Developed by Oxford University, this AI chatbot helps doctors accurately prescribe medications and flags potential adverse interactions. It aims to reduce medication errors and has performed at clinical expert levels in testing.

APPENDIX Y: SOME OF THE MOST COMMONLY USED AI TOOLS IN RETAIL AND E-COMMERCE

1. Personalization & Recommendation Engines

- **Amazon Personalize, Dynamic Yield, Algolia Recommend** – AI systems that analyze browsing and purchase data to recommend products tailored to each shopper.
- **Salesforce Einstein** – embedded in Salesforce Commerce Cloud, providing AI-driven personalization and product recommendations.

2. Customer Service & Chatbots

- **Ada, LivePerson, Drift, Zendesk AI, Intercom Fin** – conversational AI assistants for answering FAQs, resolving order issues, and offering 24/7 customer service.
- **ChatGPT (via APIs)** – increasingly embedded in custom e-commerce chat flows.

3. Search & Discovery Optimization

- **Klevu, Algolia, Bloomreach** – AI-powered search tools that improve product discovery, autocomplete, and ranking based on customer intent.
- **Google Cloud Retail Search** – designed specifically to optimize product search relevance.

4. Pricing & Demand Forecasting

- **Revionics (by Aptos)** – dynamic pricing and promotions optimization.
- **Blue Yonder, Pricemoov, Competera** – demand forecasting, revenue management, and AI-based price elasticity modeling.

5. Inventory & Supply Chain Optimization

- **Llamasoft (Coupa AI), Blue Yonder, o9 Solutions** – predictive demand planning, stock allocation, and logistics optimization.
- **Focal Systems** – AI-powered shelf monitoring with computer vision to reduce stockouts.

6. Marketing & Customer Insights

- **Klaviyo (with AI)** – personalized email and SMS marketing automation.
- **Emarsys (SAP)** – omnichannel customer engagement powered by AI.
- **Persado** – AI-generated marketing copy and emotional engagement optimization.

7. Visual Recognition & AR/VR Shopping

- **Syte, ViSenze, Google Lens API** – visual search allowing customers to upload images and find similar products.
- **Shopify AR, IKEA Place App** – AR visualization for furniture, fashion, and home goods.

8. Fraud Detection & Security

- **Signifyd, Riskified, Forter** – fraud detection platforms powered by AI to identify suspicious transactions in real time.
- **Stripe Radar** – AI-driven fraud prevention for online payments.

APPENDIX Z: SOME OF THE MOST COMMONLY USED AI TOOLS IN REAL ESTATE AND PROPERTY MANAGEMENT

1. Property Search & Valuation

- **Zillow Zestimate** – AI-driven home value estimation.
- **Redfin Estimate** – predictive property valuation using market data.
- **Reonomy** – commercial property intelligence and ownership insights.
- **HouseCanary** – valuation models and real estate analytics.

2. Predictive Analytics & Investment Platforms

- **Cherre** – real estate data integration and predictive analytics.
- **CoreLogic** – property data, risk, and valuation analytics.
- **CompStak** – AI-driven commercial lease and sales comps.
- **Mashvisor** – rental performance prediction for investors.

3. Virtual Tours & Visualization

- **Matterport** – AI-powered 3D property scans and virtual tours.
- **EyeSpy360** – 360° virtual tour creation.
- **Rooomy** – virtual staging and 3D visualization.

4. Smart Property & Facility Management

- **Building Engines (by JLL)** – AI-enabled building operations and tenant experience.
- **Facilio** – predictive maintenance and IoT-driven property operations.
- **Brivo** – AI-based smart access control.

5. Tenant Experience & Leasing Automation

- **Hello Alfred** – tenant engagement and concierge services powered by AI.
- **AppFolio AI Leasing Assistant** – automates tenant inquiries and scheduling.
- **Knock CRM** – leasing CRM with AI-driven engagement.

6. Marketing & Lead Generation

- **Zumper** – AI-driven rental listings and tenant matching.
- **RentCafe** – predictive lead scoring and automated renter outreach.
- **CRM platforms (HubSpot, Salesforce with AI plugins)** – automate marketing and lead nurturing.

7. Risk & Fraud Detection

- **LexisNexis Risk Solutions** – tenant screening and fraud prevention.
- **Experian RentBureau** – tenant credit and risk analysis.
- **Rently** – secure AI-powered self-touring for rental properties.

REFERENCES

Accenture. (2020). *Responsible AI: From principles to practice.* Accenture Insights.

Accenture. (2020). *The future of customer experience: Omnichannel is no longer optional.* Accenture Research.

Accenture. (2021). *Reinventing cost management with AI.* Accenture Strategy Report.

Accenture. (2021). *AI-powered workforce management solutions.* Accenture Insights.

Association of Certified Fraud Examiners (ACFE). (2022). *Report to the nations: Global study on occupational fraud and abuse.* ACFE.

AT&T. (2018). *AT&T's $1 billion reskilling program.* AT&T Press Release.

Bank of America. (2021). *Bank of America's Erica surpasses 1 billion client interactions.* Bank of America Newsroom.

Barocas, S., & Selbst, A. D. (2016). Big data's disparate impact. *California Law Review, 104*(3), 671–732.

Boushey, H., & Glynn, S. J. (2012). *There are significant business costs to replacing employees.* Center for American Progress.

Brynjolfsson, E., & McElheran, K. (2016). The rapid adoption of data-driven decision-making. *American Economic Review, 106*(5), 133–139.

Bughin, J., Seong, J., Manyika, J., Chui, M., & Joshi, R. (2018). *Notes from the AI frontier: Modeling the impact of AI on the world economy.* McKinsey Global Institute.

CB Insights. (2021). *Top reasons startups fail.* CB Insights Research.

Chen, L., Mislove, A., & Wilson, C. (2016). An empirical analysis of algorithmic pricing on Amazon Marketplace. *WWW Conference*, 1339–1349.

Cohen, P., Hahn, R., Hall, J., Levitt, S., & Metcalfe, R. (2016). Using big data to estimate consumer surplus: The case of Uber. *NBER Working Paper Series, 22627.*

Cokins, G. (2013). Budgeting, planning, and forecasting in uncertain times. *Journal of Corporate Accounting & Finance, 24*(3), 19–25.

Choi, T. M., Wallace, S. W., & Wang, Y. (2020). Big data analytics in operations management. *Production and Operations Management, 29*(1), 3–16; 29(6), 1221–1245;

Cisco. (2022). *Consumer privacy survey: Building consumer confidence through transparency and control.* Cisco Research.

Cowgill, B. (2019). Bias and productivity in humans and algorithms. *Labour Economics, 61*, 101747.

Deloitte. (2020). *Finance 2025: Digital transformation in finance.* Deloitte Insights.

Deloitte. (2021). *AI adoption in small and mid-sized businesses.* Deloitte Insights.

Deloitte. (2021). *The future of work in finance: Reskilling in an AI world.* Deloitte Insights.

Deloitte. (2021). *Customer analytics: Driving growth through insight.* Deloitte Insights.

Deloitte. (2022). *Supply chain resilience report: Adapting to disruption.* Deloitte Insights.

Duhigg, C. (2012). *How companies learn your secrets.* The New York Times Magazine.

Epsilon. (2018). *The power of me: The impact of personalization on marketing performance.* Epsilon Research.

European Commission. (2018). *General Data Protection Regulation (GDPR)*. Publications Office of the EU.

European Commission. (2023). *Proposal for a Regulation on Artificial Intelligence (AI Act)*. Publications Office of the EU.

Experian. (2021). *Global data management research report*. Experian Insights.

EY. (2021). *How AI is reshaping treasury and cash flow management*. Ernst & Young.

Ezrachi, A., & Stucke, M. E. (2017). Algorithmic collusion: Problems and counter-measures. *OECD Competition Committee Discussion Paper*.

Federal Trade Commission (FTC). (2021). *Aiming for truth, fairness, and equity in your company's use of AI*. FTC Business Blog.

Few, S. (2012). *Show Me the Numbers: Designing Tables and Graphs to Enlighten*. Analytics Press.

Floridi, L., & Cowls, J. (2019). A unified framework of five principles for AI in society. *Harvard Data Science Review, 1*(1).

Gallup. (2022). *State of the global workplace report*. Gallup.

Gandomi, A., & Haider, M. (2015). Beyond the hype: Big data concepts, methods, and analytics. *International Journal of Information Management, 35*(2), 137–144.

Gartner. (2019). *Why so many AI initiatives fail*. Gartner Research.

Gartner. (2021). *The state of data quality*. Gartner Research.

Gartner. (2021). *Market Guide for Augmented Analytics Tools*. Gartner Research.

Gartner. (2021). *AI in cash flow forecasting: Improving accuracy and agility*. Gartner Research.

Gartner. (2022). *Trust in AI: Building explainability in financial decision-making*. Gartner Research.

Gómez-Uribe, C. A., & Hunt, N. (2016). The Netflix recommender system: Algorithms, business value, and innovation. *ACM Transactions on Management Information Systems, 6*(4), 1–19.

Guidotti, R., Monreale, A., Ruggieri, S., Turini, F., Giannotti, F., & Pedreschi, D. (2018). A survey of methods for explaining black box models. *ACM Computing Surveys, 51*(5), 93.

Harver. (2021). *How Unilever uses AI in recruitment*. Harver Case Study.

IBM. (2019). *How IBM HR uses AI to reduce turnover*. IBM Research.

IBM. (2023). *Cost of a data breach report*. IBM Security.

IDC. (2022). *DataSphere and StorageSphere Forecast 2022–2026*. IDC Research.

Intuit. (2021). *QuickBooks introduces AI-driven cash flow forecasting*. Intuit Press Release.

Ivanov, D., & Dolgui, A. (2020). Viability of intertwined supply networks: extending the supply chain resilience angles towards survivability. *International Journal of Production Research, 58*(10), 2904–2915.

JPMorgan Chase. (2017). *How COiN is transforming contract review*. JPMorgan Chase & Co.

JPMorgan Chase. (2017). *How JPMorgan is using AI to review legal documents*. JPMorgan Insights.

Juniper Research. (2020). *Chatbots: Banking, eCommerce, Retail & Healthcare 2020–2024*. Juniper Insights.

Kashyap, R., & Raghu, T. S. (2018). Transforming workforce scheduling with analytics: The case of retail services. *MIS Quarterly Executive, 17*(2), 127–144.

Khatri, V., & Brown, C. V. (2010). Designing data governance. *Communications of the ACM, 53*(1), 148–152.

Klumpp, M. (2018). Automation and artificial intelligence in business logistics systems. *Business and Information Systems Engineering, 60*(4), 269–280.

Lam, H., Kavanagh, M., & Serpa, R. (2015). The impact of scheduling practices on retail employees. *Journal of Business and Psychology, 30*(2), 421–434.

Lemon, K. N., & Verhoef, P. C. (2016). Understanding customer experience throughout the customer journey. *Journal of Marketing, 80*(6), 69–96.

LinkedIn. (2022). *Global talent trends*. LinkedIn Report.

Mandinach, E. B., & Gummer, E. S. (2016). What does it mean for teachers to be data literate? *Journal of Educational Research and Policy Studies, 16*(1), 12–34.

Manyika, J., et al. (2017). *Jobs lost, jobs gained: Workforce transitions in a time of automation.*

Martin, K., & Murphy, P. (2017). The role of data privacy in marketing. *Journal of the Academy of Marketing Science, 45*(2), 135–155.

Medhat, W., Hassan, A., & Korashy, H. (2014). Sentiment analysis algorithms and applications: A survey. *Ain Shams Engineering Journal, 5*(4), 1093–1113.

Microsoft. (2021). *How Microsoft uses AI to transform finance*. Microsoft Case Study.

McKinsey & Company. (2013). *Big data: The next frontier for innovation, competition, and productivity.*

McKinsey & Company. (2018). *The future of pricing: How big data and AI are changing the game.*

McKinsey & Company. (2018). *Smartening up with artificial intelligence: What's in it for industry?*

McKinsey & Company. (2019). *The value of customer experience: How companies can grow by making CX a priority.*

McKinsey & Company. (2020). *The future of forecasting.*

McKinsey & Company. (2020). *Risk, resilience, and rebalancing in global value chains.*

McKinsey & Company. (2021). *Next in personalization 2021 report.*

McKinsey & Company. (2021). *The state of AI in 2021.*

Mehrabi, N., Morstatter, F., Saxena, N., Lerman, K., & Galstyan, A. (2021). A survey on bias and fairness in machine learning. *ACM Computing Surveys, 54*(6), 1–35.

New York Times. (2017). *After Hurricane Harvey, price gouging complaints surge.*

Ngai, E. W. T., Hu, Y., Wong, Y. H., Chen, Y., & Sun, X. (2011). Data mining techniques in financial fraud detection. *Decision Support Systems, 50*(3), 559–569.

Ngai, E. W. T., Hu, Y., Wong, Y. H., Chen, Y., & Sun, X. (2011). The application of data mining techniques in financial fraud detection. *Decision Support Systems, 50*(3), 559–569.

NewVantage Partners. (2022). *Data and AI leadership executive survey*. NewVantage Partners Report.

Obermeyer, Z., Powers, B., Vogeli, C., & Mullainathan, S. (2019). Dissecting racial bias in an algorithm used to manage population health. *Science, 366*(6464), 447–453.

Otto, B. (2011). Organizing data governance. *Communications of the Association for Information Systems, 29*(1), 45.

Palmer, A. (2021). Amazon's automated management of warehouse workers. *CNBC Tech Report.*

Park, Y., Hong, P., & Roh, J. J. (2013). Supply chain lessons from the catastrophic natural disaster in Japan. *Business Horizons, 56*(1), 75–85.

PwC. (2020). *AI in healthcare workforce management*. PwC Insights.

PwC. (2020). *Cash analytics and AI: The future of liquidity management.* PwC.

PwC. (2018). *Future of customer experience survey.* PwC Research Report.

PwC. (2021). *Consumer intelligence series: The growth of sustainability.* PwC Research Report.

PwC. (2021). *Consumer trust insights survey.* PwC Research Report.

PwC. (2021). *CFO survey: The future of finance.* PwC.

Ransbotham, S., Khodabandeh, S., Fehling, R., LaFountain, B., & Kiron, D. (2021). Achieving individual—and organizational—value with AI. *MIT Sloan Management Review, 63*(1), 1–13.

Rawson, A., Duncan, E., & Jones, C. (2013). The truth about customer experience. *Harvard Business Review, 91*(9).

Redman, T. C. (2018). Seizing opportunity in data quality. Harvard Business Review, 96(4), 88–97.

Rodrigues, A. M., Stantchev, D., Potter, A., Naim, M. M., & Whiteing, A. (2018). Establishing a transport operation performance measurement framework. *International Journal of Logistics Management, 19*(2), 205–225.

Sephora. (2021). *How Sephora leverages AI for personalized and omnichannel beauty experiences.* Sephora Corporate Insights.

Shamim, S., Cang, S., Yu, H., & Li, Y. (2019). Examining the feasibility of sentiment analysis approaches in HRM. *International Journal of Information Management, 46,* 139–150.

SHRM. (2022). *State of the workplace report.* Society for Human Resource Management.

Shrestha, Y. R., Ben-Menahem, S. M., & von Krogh, G. (2019). Organizational decision-making structures in the age of AI. *California Management Review, 61*(4), 66–83.

Siemens. (2021). *How Siemens uses AI in smart factories.* Siemens Industry White Paper.

Sivarajah, U., Kamal, M. M., Irani, Z., & Weerakkody, V. (2017). Critical analysis of big data challenges. *Journal of Business Research, 70,* 263–286.

Smith, B., & Linden, G. (2017). Two decades of recommender systems at Amazon.com. *IEEE Internet Computing, 21*(3), 12–18.

Starbucks. (2019). *How Starbucks uses Deep Brew AI to personalize customer experiences.* Starbucks Investor Relations.

Tableau. (2021). *AI-powered features: Explain Data and Ask Data.* Tableau Product Documentation.

Tseng, M. L., Tan, R. R., Chiu, A. S., Chien, C. F., & Kuo, T. C. (2019). Circular economy meets industry 4.0: Can big data drive industrial sustainability? *Resources, Conservation and Recycling, 151,* 104482.

UK Competition and Markets Authority (CMA). (2021). *Algorithms: How they can reduce competition and harm consumers.* CMA Report.

Unilever. (2021). *Driving sustainable growth through digital finance.* Unilever Annual Report.

Unilever. (2022). *Unilever sustainable living plan: Progress and insights.* Unilever Corporate Report.

UPS. (2020). *UPS ORION: Transforming logistics with AI route optimization.* UPS Corporate Sustainability Report.

Voigt, P., & Von dem Bussche, A. (2017). *The EU General Data Protection Regulation (GDPR): A practical guide.* Springer.

West, J., & Bhattacharya, M. (2016). Intelligent financial fraud detection: A comprehensive review. *Computers & Security, 57,* 47–66.

Willcocks, L. P. (2020). Robotic process automation and the future of work. *Journal of Information Technology, 35*(2), 124–135.

World Economic Forum. (2021). *How supply chains rose to the COVID-19 challenge.* World Economic Forum Report.

World Economic Forum (WEF). (2022). *AI democratization: Leveling the playing field for small businesses.* WEF Report.

Zhang, Y., et al. (2021). Algorithmic management in the workplace: Implications for HR. *Human Resource Management Review, 31*(4), 100765.

Zhou, Z.-H., et al. (2017). Weakly supervised learning. *National Science Review, 5*(1), 44–53.

Zillow. (2019). *Zestimate accuracy fact sheet.* Zillow Research.

www.ingramcontent.com/pod-product-compliance
Lightning Source LLC
Chambersburg PA
CBHW060326200326
41519CB00011BA/1849